Geometry of the Fundamental Interactions

M.D. Maia

Geometry of the Fundamental Interactions

On Riemann's Legacy to High Energy Physics and Cosmology

M.D. Maia
Universidade de Brasilia
Institute of Physics
70910-000 Brasilia D.F.
Brazil
maia@unb.br

ISBN 978-1-4899-9024-2 ISBN 978-1-4419-8273-5 (eBook)
DOI 10.1007/978-1-4419-8273-5
Springer New York Dordrecht Heidelberg London

Softcover re-print of the Hardcover 1st edition 2011

Artwork by Diego Moscardini

Printed on acid-free paper

Springer is part of Springer Science+Business Media (www.springer.com)

Preface

The four fundamental forces in nature, gravitation, electromagnetic, weak, and strong nuclear forces, are based on a single idea of the 19th century, the Riemann curvature. The vast amount of experimental data and theoretical development in high energy physics has confirmed that concept. Only very recently, Einstein's gravitational field, which originated the geometric paradigm for physics, has shown signs that it needs an improvement to explain the gravitational observations in modern cosmology, where Einstein's gravitational field can describe only about 4% of the gravitational interaction in the universe. On the other hand, at the quantum scale Einstein's gravitational field has resisted all attempts to quantization. Therefore, something appears to be missing to complete the idea of Riemann.

In the past 20 years we have debated with colleagues, teachers, collaborators, and students on the different forms in which geometry and the physics of the fundamental interactions mix. The overall feeling is that the understanding of the geometry of the fundamental interactions has become too complex to grasp within the standard professional lifetime of a graduate student of physics, mathematics, astronomy, and engineering to understand what is going on, specially within the current productivity syndrome. Hence the proposal of this book to supply a blend of what is known and what is not explained.

Therefore, the program of this book is about theoretical research with emphasis on inducing a debate, whenever possible, on how to fix and improve existing theories which have reached their applicability and prediction limits. We start with concepts of physical space since Kant, going through the evolution of the idea of space–time, symmetries and its associated connections, the Yang–Mills theory, and ending with gravitation, including a conceptual discussion on the deficiencies of Riemann curvature, which is the central theme of the book.

The author wishes to thank the many contributions resulting from classroom and coffee break debates during the years when we have lectured on the subject. He also thanks the suggestions and comments from colleagues of the Mathematics and Physics departments on earlier drafts, from which much was learned.

Brasilia, Brazil
December 2010

M.D. Maia

Contents

1 The Fundamental Interactions ... 1

2 The Physical Manifold ... 9
- 2.1 Manifolds ... 9
- 2.2 Geometry of Manifolds ... 18
- 2.3 The Riemann Curvature ... 21

3 Symmetry ... 25
- 3.1 Groups and Subgroups ... 25
- 3.2 Groups of Transformations ... 27
- 3.3 Lie Groups ... 29
- 3.4 Lie Algebras ... 31
  - 3.4.1 Infinitesimal Coordinate Transformations ... 31
  - 3.4.2 Infinitesimal Transformations on Vector Bundles ... 33

4 The Algebra of Observables ... 43
- 4.1 Linear Form Fields ... 44
- 4.2 Tensors ... 48
- 4.3 Exterior Algebra ... 50

5 Geometry of Space–Times ... 57
- 5.1 Galilean Space–Time ... 57
- 5.2 Newton's Space–Time ... 61
  - 5.2.1 The Curvature of Newton's Space–Time ... 62
- 5.3 The Minkowski Space–Time ... 64
- 5.4 Space–Times in General Relativity ... 67

6 Scalar Fields ... 73
- 6.1 Classic Scalar Fields ... 77
- 6.2 Non-linear Scalar Fields ... 82

7 Vector, Tensor, and Spinor Fields 89
7.1 Vector Fields 89
7.1.1 The Electromagnetic Field 89
7.1.2 The Maxwell Tensor 93
7.1.3 The Nielsen–Olesen Model 96
7.2 Spinor Fields 100
7.2.1 Spinor Transformations 104

8 Noether's Theorem 107
8.1 Noether's Theorem for Coordinate Symmetry 107
8.2 Noether's Theorem for Gauge Symmetries 114

9 Bundles and Connections 125
9.1 Fiber Bundles 125
9.2 Base Morphisms 127
9.3 Principal Fiber Bundles 130
9.4 Connections 132

10 Gauge Fields 139
10.1 Gauge Curvature 139
10.2 The $U(1)$ Gauge Field 142
10.3 The $SU(2)$ Gauge Field 146
10.4 The $SU(3)$ Gauge Field 153

11 Gravitation 157
11.1 The Riemann Curvature 157
11.2 Gauge Gravity 159
11.3 Loop Gravity 163
11.4 Deformable Gravity 164
11.5 Kaluza–Klein Gravity 169

References 171

Index 177

List of Figures

1.1 The Aharonov–Bohm experiment ... 5
2.1 Manifold ... 11
2.2 Manifold mappings ... 12
2.3 Curve on a manifold ... 13
2.4 The tangent bundle ... 15
2.5 The Riemann curvature ... 21
4.1 The derivative map and its dual ... 47
5.1 The Galilean space–time $\mathscr{G}_4$ ... 59
5.2 Projecting the Galilean space–time on $I\!R^3$ ... 59
5.3 The Newtonian space–time ... 62
5.4 Geodesic deviation in $\mathscr{N}_4$ ... 63
6.1 Quartic potential ... 83
6.2 The Mexican hat quartic potential for two scalar fields ... 86
8.1 Map between fibers from a coordinate transformation ... 108
8.2 Boundaries of a closed region in $\mathscr{M}$ ... 114
8.3 Gauge transformation ... 116
9.1 Cylinder fiber bundle ... 126
9.2 The Möbius fiber bundle ... 126
9.3 Trivialization of a fiber bundle ... 128
9.4 Compatible trivializations ... 129
9.5 Correspondence between X^a and dx^μ ... 135

Chapter 1
The Fundamental Interactions

The four fundamental interactions that we recognize today are the *gravitational, electromagnetic, weak and strong nuclear forces*. So far they have been sufficient to describe most of the observed properties of matter, but it is not impossible that some other force will assume such fundamental role in the future. However, not everyone agrees that there are compelling evidences for such assumption, at least within a simple and understandable argument in the Occam sense.[1] Indeed, recent cosmological evidences show that the remaining 96% of the known universe may be filled with something that we do not quite understand, and this is very appropriately called the dark component of the universe, composed of dark matter showing attractive gravitation and dark energy showing something like a repulsive gravitation. They are dark because they cannot be seen, like ordinary matter.

The missing matter in galaxies and clusters was noted in 1933 by Fritz Zwicky, when he was observing the rotation of stars in galaxies using Newton's gravitational law. The name dark matter is credited to Vera Rubin in 1970 [1, 2]. Dark energy is far more recent, appearing in 1998 as a possible explanation for the accelerated expansion of the universe, observed by search teams looking at very distant type Ia supernovae [3]. An alternative current thinking about dark energy is that it is the quantum vacuum, something like the void proposed by Thomas Bradwardine in another dark age [4], but a void endowed with some energy. Such new cosmology is an integrated science involving not only optical ground-based and space telescopes and radio telescopes operating in a wide rage of frequencies up to x-rays but also gravitational wave detectors, cosmic ray detectors, deep underground neutrino experiments and high energy particle colliders, and geometry and mathematical analysis.

On the theoretical side of these fascinating and hard to ignore facts, there is a strong competition between explanations, not all of them following Occam's maxim or the existence of a classical or quantum void. The ideas vary from the supposed existence of new and exotic forms of matter; or that space–time may have more

[1] The maxim that the true explanation for some natural phenomenon is also the simplest one is generally attributed to William de Occam, a Franciscan friar and philosopher from the middle ages period.

M.D. Maia, *Geometry of the Fundamental Interactions*,
DOI 10.1007/978-1-4419-8273-5_1,

than the usual four dimensions; or that space–time may have a foam-like structure formed by "atoms" of space–time; or that space–time is as smooth as possible; or that the universe exists because we are here; or several combinations of these. Here we present just a glimpse of these concepts that are part of well-established theories. We will not discuss other theories which are not proven consistent either theoretically or experimentally. We start at the beginning, with the ideas of geometry and topology of Riemann which emerged around 1850 [5, 6].

Since the advent of Riemann's concepts of curvature of a manifold, it has become the main tool behind all the modern theories of fundamental interactions. In the following sections we will discuss the essential ideas of Riemann, Lie, Weyl, Einstein, and others and how they have established such stronghold for modern science.

As we shall see, the Riemann curvature depends on the preliminary notion of an *affine connection* or equivalently of a covariant derivative. On the other hand, as a consequence of Noether's theorem, the curvature appears as observables in nature, responsible for the fundamental interactions. Therefore, we may pause for a reflection on the possibility that geometry and analysis are ultimately dependent on the underlying physics of the fundamental interactions [7].

Einstein's general relativity of 1916 set the geometrical paradigm that we use today based on the notion of curvature set by Riemann. Thus, we no longer think of gravitation as a force, but rather as a curvature of space–time, as compared with the idealized Minkowski's flat space–time. It was only after 1954, with the works of C. N. Yang and R. Mills, that it was understood that the other three fundamental interactions, known as *gauge interactions*, also have the same curvature meaning. However, this latter development is not intuitive and it took a long time to mature [8].

The development of gauge theory started in 1919, with two independent ideas. The first one was the proposal of Hermann Weyl to describe a geometrical theory of the electromagnetic field [9]. The second one was the development by Emmy Noether of very general theorems concerning the construction of the observables of a physical theory, starting from the knowledge of its symmetries [10].

Weyl's original idea was to generalize Einstein's gravitational theory by incorporating the electromagnetic field as part of the space–time geometry: He reasoned that in the same way as the gravitational field is defined by the quadratic form defined in space–time

$$ds^2 = g_{\mu\nu} dx^\mu dx^\nu$$

where the coefficients $g_{\mu\nu}$ are solutions of Einstein's equations, the electromagnetic field would be defined by the coefficients of a linear form

$$dA = A_\mu dx^\mu$$

where A_μ are the components of the electromagnetic four-vector potential. To achieve such geometric unification of gravitation and electromagnetism, Weyl modified the metric affine connection condition $g_{\mu\nu;\rho} = 0$ (the so-called *metricity condition* of Riemann's geometry), where the semicolon denotes Riemann's covariant derivative, by the more general condition [9]

$$g_{\mu\nu;\rho} = -2A_\rho g_{\mu\nu} \tag{1.1}$$

In this way Weyl hoped to obtain the electromagnetic field described by the Maxwell tensor

$$F_{\mu\nu} = \frac{\partial A_\mu}{\partial x^\nu} - \frac{\partial A_\nu}{\partial x^\mu}$$

satisfying Maxwell's equations

$$F^{\mu\nu}{}_{;\nu} = J^\mu \tag{1.2}$$

$$F_{\mu\nu;\rho} + F_{\rho\mu;\nu} + F_{\nu\rho;\mu} = 0 \tag{1.3}$$

where the partial derivatives (,) are replaced by the covariant derivative (;) defined by Weyl's connection satisfying condition (1.1) and where J^μ are the components of the four-dimensional current density (see e.g., [11]).

Weyl's proposal did not succeed essentially because in translating Maxwell's equations to his new geometry, the Poincaré symmetry was lost and so also the compatibility with the gauge transformations of the electromagnetic potential. Indeed, the above expressions for $F_{\mu\nu}$ are covariant (that is, they keep the same form on both sides) under the Poincaré transformations in space–time and also under the transformations of the electromagnetic potential given by

$$A'_\mu = A_\mu + \partial_\mu \theta(x) \tag{1.4}$$

where, as indicated, the parameter $\theta(x^\mu)$ is a function of the space–time coordinates. These transformations are not a consequence of the Poincaré transformations of coordinates, but they form a group by their own properties, acting in the space of the potentials, in such a way that they are compatible with the Poincaré coordinate symmetry. Therefore, when Weyl tried to write the gauge transformations (1.4) of the electromagnetic field in a curved space–time, the Poincaré symmetry was replaced by the group of diffeomorphisms of coordinate transformations of the curved space–time. In doing so, the mentioned compatibility between the two symmetries was lost. This result implied that at each point of the curved space–time, the potential A_μ would not behave as the known electromagnetic potential (which can be calibrated by all observers in the orderly way dictated by (1.4)). In the Weyl proposal, the two symmetries would lead to unpredictable results for the electromagnetic field. In view of such inconsistency, Weyl abandoned his theory.

The solution to *Weyl's gauge inconsistency* appeared only after the development of quantum theory. In 1927 Vladmir Fock and Fritz London suggested that Weyl's idea could in principle make sense in quantum mechanics, where the gauge transformation (1.4) would be replaced by a *unitary gauge transformation* like [12, 13]

$$A'_\mu = e^{i\Theta(x)} A_\mu$$

Since in quantum theory only the norm $||A_\mu||$ is an observable, the above unitary transformation would not affect the observable potentials themselves, regardless of the type of coordinate transformations. Consequently, the Fock–London suggestion would apply only to a quantum version of the electromagnetic theory, with the above unitary gauge transformations.

With this new interpretation, Weyl reconsidered his theory in 1929, when he introduced the concept of *gauge transformations* in the sense that it is used today, specially including the "local gauge transformations" in which the parameters are dependent on coordinates. Such unitary gauge transformation would be understood as an intrinsic property of the quantum electromagnetic field, retaining its Poincaré invariance in the classical limit [14]. In this case, the diffeomorphism invariance of his theory would not interfere with the unitary transformations of that quantum gauge transformations.

However, little was known in 1929 about the quantum behavior of electrodynamics. With the lack of experimental support to the quantum interpretation, Weyl's theory entered into a second dormant period lasting to about 1945, when new properties of fields and elementary particles would become more evident. The resulting theory called *quantum electrodynamics* (or QED) associated with the one-parameter unitary gauge group $U(1)$ was described by J. Schwinger around 1951 [15]. Instead of describing just the already known classical electromagnetic interaction between charged particles, in QED the interaction between charged particles was intermediated by photons.

The second important contribution to gauge theory from the period 1918 to 1919 was the theorem by Emmy Noether showing how to construct the observables of a physical theory, starting from the knowledge of its Lagrangian and its symmetries [10]. Although the motivations and results of Weyl and Noether were independent, they met at the point where Noether introduced a matrix-vector quantity (a vector whose components are matrices) to obtain the divergence theorem in the case where the parameters depended on the coordinates. Later on it was understood that Noether's matrix-vector defined the same gauge potential when it is written in the *adjoint representation* of the *Lie algebra* of the gauge group. Then the properties of the adjoint representations of the Lie algebras of local symmetry groups became central to the development of gauge theory. More than that, Noether's theorem made it possible to predict new results from gauge theories. It also meant that the gauge potentials are observables and not just mathematical corrections to derivatives and to the divergence theorem.

In 1956 Yakir Aharonov and David Bohm suggested an experiment to find an observable effect associated with the magnetic potential vector **A** (and not by the magnetic field **B**) itself, on the electron deviation in a double slot experiment. The proposed experiment used a long spiral coil parallel to the slots, in such a way that the magnetic field being confined to the center of the coil would not interfere directly on the electron beam [16]: The experiment was realized in 1960, showing that indeed there was shift on the phase of the wave function, given explicitly by the vector potential

$$\Psi' = e^{-\frac{ie}{\hbar}\oint_c <\mathbf{A},d\ell>}\Psi \tag{1.5}$$

where the integral is calculated on the closed path c formed by the electron beams in Fig. 1.1. Since this is essentially the consequence of a phase transformation depending on the local coordinates, this experiment can also be seen as a confirmation of the Fock and London interpretation.

The rest of the history of development of gauge theory represents an exuberant mixture of theory and experiment, mainly because it depended on the development of nuclear theory and indeed on the whole phenomenology of particle physics. The reader may find details in, e.g., [17–19] among other fine reviews.

In 1932 Werner Heisenberg had already proposed that protons and neutrons could be described as being distinct states of one same particle, the nucleon [20]. This nucleon was consistently described by a new quantum number, the isotopic spin (or *isospin*), mathematically described by the $SU(2)$ group. This is formally similar to the orbital spin, but here it is regarded as an *internal symmetry*. However, a major difference with the QED gauge theory is that the $SU(2)$ isospin symmetry is a *global symmetry*, in the sense that its parameters do not depend of the coordinates [21].

In 1954 Yang and Mills proposed a generalization of electrodynamics, where the $U(1)$ group was replaced by the *local* $SU(2)$ group. In this case, instead of a vector potential A_μ, the proposed theory has a non-Abelian 2×2 matrix-vector potential $\mathbf{A}_\mu$, whose components are defined in the Lie algebra of the new symmetry [22], similar to the matrix-vector invented by Noether. However, the physical interpretation of the local $SU(2)$ required further developments, like the Weinberg–Salam theory.

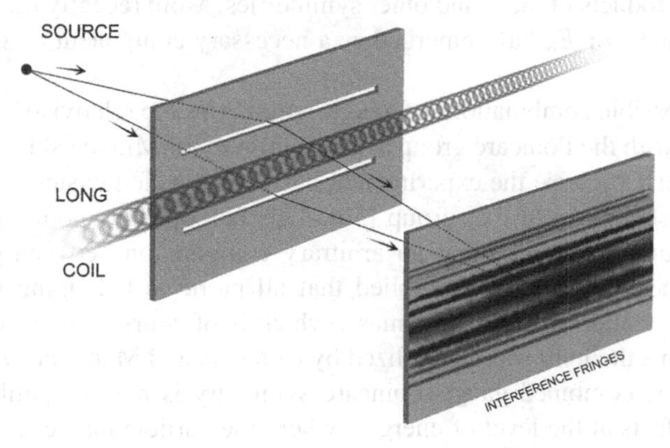

Fig. 1.1 The Aharonov–Bohm experiment

In the period 1967–1968, Weinberg and Salam independently proposed a unification of the electrodynamics with the weak nuclear force called the *electroweak theory* based on the symmetry $SU(2) \times U(1)$ [23–25]. This theory predicted three intermediate particles with integer spins, the W^+, W^-, and Z^0 bosons, whose existence was confirmed experimentally at CERN in 1983, producing also the experimental evidence of the $SU(2)$ gauge theory.

The existence of sub-nuclear particles called quarks (the word quark was extracted from Finnegan's Wake [26] by Gell-Mann in 1964). In 1961 Gell-Mann and independently Yuval Ne'eman had formulated a particle classification scheme called the eightfold way, which after much hard work resumed in the $SU(3)$ symmetry [27, 28]. In this scheme quarks were held together by gluons in a more general model of strong nuclear interactions, today understood as a Yang–Mills theory with eight local parameters organized in a local $SU(3)$ gauge symmetry. The $SU(3)$ group contains $SU(2)$ and $U(1)$ as subgroups associated with the isospin and hypercharge, respectively. The four remaining parameters would describe the components of the strong nuclear force called gluons, represented in the Lie algebra of $SU(3)$ as the bounding force between quarks. The result is a theory of strong interactions nicknamed quantum chromodynamics (QCD). At the time of this writing, quarks were never observed as free particles but always bounded to another by gluons. Summarizing, the original works of Weyl and Noether took shape with the name of gauge field theory (or Yang–Mills theory), involving three of the four fundamental interactions, associated with the local gauge symmetries $U(1)$, $SU(2)$, and $SU(3)$, respectively.

Those three groups can be seen as parts or subgroups of a larger symmetry group, *the combined symmetry group*. The simplest combined symmetry is just the Cartesian product $U(1) \times SU(2) \times SU(3)$, which forms the basic or *standard model* of particle interactions. It was soon found that the standard model is not sufficient to describe other aspects of the structure of particle physics, like, for example, their organization into families. Thus, a more general group of symmetries was and still is sought for, which can eventually lead to a grand unification theory (GUT), involving the three gauge interactions. Suggested candidates are $SU(5)$, $SU(6)$, $SO(10)$, and products of these and other symmetries. More recently the exceptional groups such as E_7 or E_8 have emerged as a necessary component of such scheme [29].

Since all possible combinations of gauge symmetries are relativistic, they should also combine with the Poincaré group as the symmetry of Minkowski's space–time. This is necessary because the experimental basis of particle physics is constructed with the representations of that group [30]. This fact opened another problem: In 1964 O'Raifeartaigh showed that an arbitrary combination between gauge symmetries and the Poincaré group implied that all particles belonging to the same spin multiplet would have the same mass, which is of course not correct. In 1967 O'Raifeartaigh's theorem was generalized by Coleman and Mandula, with the same conclusion: The combined gauge–Poincaré symmetry is not compatible with the experimental facts at the level of energies where the particle masses are evaluated.

This mass splitting problem became known as the *no-go theorem* for the compatibility between particle physics and gauge theory.

A more detailed analysis of these theorems shows that the difficulty lies in the translational subgroup of the Poincaré group. The conclusion is that either the Poincaré translations are left aside, or else the Lie algebra structure should change, or finally that the combined symmetry would not hold at the level of measurement of the particle masses [31–33]. Clearly, such fundamental issues required a radical solution if the whole scheme of gauge theories was to succeed.

Among these proposed solutions, one suggested the replacement of the Poincaré group by the deSitter group, which have the same number of parameters as the Poincaré group [34]. This choice would be naturally justified by the presence of the cosmological constant in Einstein's equations, which forbids the Poincaré symmetry in favor of the deSitter group. The currently observed acceleration of the universe finds in the cosmological constant a simple explanation, provided the cosmological constant problem can be explained.

In 1972 Roger Penrose, with different motivations, suggested that the translational symmetry could be hidden by use of the conformal group which is also a symmetry of Maxwell's equations [35]. The violation of the causality was perhaps the main restriction imposed on the use of the conformal group as a fundamental symmetry of physics. Further considerations on the conformal symmetry emerged again in 1998 for a possible mechanism to conciliate particle physics and gravitation. This was codenamed the ADS/CFT correspondence: Conformal invariant field theories can set in correspondence with isometric invariant fields in the five-dimensional anti-deSitter space. Since all known gauge fields are quantized, the ADS/CFT correspondence can be used to define quantum theory in a gravitational background [36] sometimes together with supersymmetric theories. Supersymmetry was introduced in 1974 by J. Wess and B. Zumino, as a modification of the Poincaré Lie algebra structure such that particles with half and integer spins would be interchangeable [37]. Since the generators of infinitesimal transformations of this new symmetry do not close as a Lie algebra, the resulting "graded Lie algebra" in principle would solve the mass splitting problem.

However, we cannot give up the Poincaré symmetry at the cost of having to define a new particle physics, based either on the deSitter group or on a supersymmetric group. As it was found later on, supersymmetry has to be broken at the lower levels of energies where the standard model of particle interactions applies. To the present, none of the new particles predicted by supersymmetric theory were found.

The presently adopted option to solve the mentioned no-go theorem is the so-called Higgs mechanism for spontaneous breaking of symmetries, proposed by Higgs in 1964. Essentially, the Higgs field is a postulated scalar field required to break the combined symmetry, so that the observed masses of the particle multiplets become distinct [38, 39].

Gravitation remains a great mystery. Assuming that the standard theory of gravitation is Einstein's general relativity, it still cannot explain the motion of stars in galaxies and clusters; the early inflation and the currently observed accelerated expansion of the universe. It has also resisted all attempts to be compatible with quantum mechanics even after nearly a century of hard work on its quantization.

The earliest consideration on quantum gravity was made by Planck in 1907, when he attempted to define a natural system of physical units, in which Newton's gravitational constant G, the speed of light c, and Planck's reduced constant $\hbar$ have value 1, and everything else would be measured in centimeters. The result is that quantum gravity would exist only at the energy level of 10^{19} Gevs, at the small length of 10^{-33} cm, which defines the so-called *Planck regime* [40]. This regime created the *hierarchy problem of the fundamental interactions*, because all other fundamental interactions exist at $\approx 10^3$ GeVs and nothing happens in between these limits. Another problem associated with the Planck regime is that it holds only in the border between three theories, Newtonian gravity, special relativity, and general relativity, each one with different symmetries and therefore with different observers. As a way to maintain a special system of units, the physics at such triple border is difficult to understand. Yet, there are over 20 theories of quantum gravity proposed up to the present ranging from the ADM program from the early 1960s, to string theory and loop quantum gravity, to massive gravity, all depending on the Planck regime.

In 1971 'tHooft showed that all gauge theories are finite when quantized by perturbative processes [41]. Therefore, if gravitation could be written as a gauge theory, then in principle we could apply 'tHooft's result to obtain a quantum theory of gravity. However, in spite of the many efforts made to write a gauge theory of gravity, it is not yet clear what is the appropriate gravitational gauge symmetry. We will return to this topic in the last chapter.

We can say that the history of the fundamental interactions is a monument to the human effort to understand nature, written in a rich mathematical language. Our objective is to discuss the various concepts involved, such as manifolds, space–times, basic field theory, symmetry, Noether's theorem, connections, culminating with our central theme, *the Riemann curvature*.

Chapter 2
The Physical Manifold

2.1 Manifolds

The basic concept of a physical space was formulated by Kant in his Critique of Pure reason 1781, where he used the word *mannigfaltigkeit* to describe the set of all *space and time perceptions* [42]. Except for the lack of specification of a geometry and of the measurement conditions, Kant's concept of physical space is very close to our present notion of space–time.

The same word *mannigfaltigkeit* was used by Riemann in 1854, with a slightly different meaning to define his metric geometry. Riemann was less emphatic on the observational detail and more concerned with the geometry itself, the idea of proximity of the objects, and with the notion of the shape or topological qualities. These concepts were introduced by Riemann in his original paper [5]. Since Riemann's paper used very little mathematical language and expressions, it led to different interpretations. The impact of that paper on essentially all modern physics, geometry, mathematical analysis, and the subsequent technology, we can hardly avoid commenting on some fundamental aspects of Riemann's geometry and how it is used today.

Riemann's paper was translated to English in 1871 by Clifford where the word *mannigfaltigkeit* was translated to "manifold," and this was subsequently adopted as the translation of *mannigfaltigkeit* in all current dictionaries. Inevitably, in the translation process, some of the original concepts of Kant, specially the perception aspect, was shaded by the concept of topological space, another invention of Riemann in the same paper [5, 43, 44].

The *topological space* of Riemann is the same as we understand today: Any set endowed with a collection of *open sets* such that their intersections and unions are also open sets and that such collection covers the whole manifold. With such topology we may define the notions of limits and derivatives of functions on manifolds [44].

Such topology is *primarily borrowed* from the metric topology of the parameter space $I\!R^n$, so that the standard mathematical analysis in Euclidean spaces can be readily used [43, 45–48]. Once this choice is made, then it is possible to define

M.D. Maia, *Geometry of the Fundamental Interactions*,
DOI 10.1007/978-1-4419-8273-5_2, © Springer Science+Business Media, LLC 2011

other topological basis, although they are not always practical as the borrowed topology of $I\!R^n$. One drawback of the borrowed topology is that a manifold can be described as being locally equivalent to $I\!R^n$, leading to the wrong interpretation that the manifold is composed of dimensionless points, like those of the $I\!R^n$. This conflicts with the Kant description of manifolds as a set of perceptions, unless we understand that point particles are not really points but just a mathematical name, capable of carrying physical qualities such as mass, charge, energy, and momenta, thus occupying a non-zero volume. In this sense a point particle can be a galaxy, an elephant, a membrane, a string, or a quark, as long as it can be assigned a time and position (as if endowed with a global positioning system (GPS)). Thus, the local equivalence between a manifold and the parameter space $I\!R^n$ does not extend to the physical meaning of the manifold. Here and in the following we use the concept of manifold as a physical space (in the sense of Kant) and often refer to its objects as points, not to be confused with the points of the parameter space.

Another topic on manifolds which deserves a comment is the choice of $I\!R^n$ as the parameter space. For some, the physical space is composed primarily of *elementary particles* and as such they should be parameterized by a discrete set and not continuous because particles are of quantum nature, characterized by a discrete spectra of eigenvalues. It is also argued that the differentiable nature associated with Riemann's topology of open sets can be replaced by a discrete topology. Thus, the usual differential equations are replaced by finite difference equations. In this interpretation the continuum would be only a non-fundamental short sight view of a discrete physical space [49–52].

On the other hand, the choice of $I\!R^n$ as the parameter space makes sense when we consider that the observers, the observables, and the conditions of measurement are defined primarily by classical observers using classical physics based on the continuum. After all, it was the differentiable structure that allowed those classical observers and their instruments to construct quantum mechanics, the present notion of elementary particles and their observables, defined by the eigenvalues of the Casimir operators of the Poincaré group. One of the most complete discussions on this fundamental subject was presented by Weyl, when he combines the foundations of mathematics with that of physics [53, 54]. In this book we base our arguments on the type of spectra of the Casimir operators. We do not see why the discrete spin spectrum of eigenvalues should be favored in presence of the spectrum of the mass operator of the Poincaré group, which, unlike the spin spectrum, is continuous (although assuming only discrete values) [31, 33]. In this sense we agree with Weyl's conclusion that the parameter space is $I\!R^n$, where continuous fields gives the fundamental physical structures with the quantum masses, spins, color, strangeness, etc. as secondary characteristics.

After these considerations we may proceed with the standard definition and properties of manifolds as found in most textbooks:

Definition 2.1 (Manifold) A manifold $\mathscr{M}$ is a set of objects (generally called points and denoted by p) with the following properties:

(a) For each of these objects we may associate n coordinates in $I\!R^n$, by means of an 1:1 map $\sigma : \mathcal{M} \rightarrow I\!R^n$,

$$\sigma(p) = (x^1, x^2, \ldots, x^n)$$

with inverse $\sigma^{-1} : I\!R^n \rightarrow \mathcal{M}$ such that

$$\sigma^{-1}((x^1, x^2, \ldots, x^n)) = p$$

(b) Given another such map τ, associate with the same p another set of coordinates $\tau : \mathcal{M} \rightarrow I\!R^n$,

$$\tau(p) = (x'^1, x'^2, \ldots, x'^n)$$

with inverse

$$\tau^{-1}((x'^1, x'^2, \ldots, x'^n)) = p$$

Then the composition $\phi = \sigma^{-1} \circ \tau : I\!R^n \rightarrow I\!R^n$ is the same as a coordinate transformation in $I\!R^n : x'^i = \phi^i(x^j)$ (see Fig. 2.1).

(c) For all points of $\mathcal{M}$ we can define one such map and the set of such maps covers the whole $\mathcal{M}$.

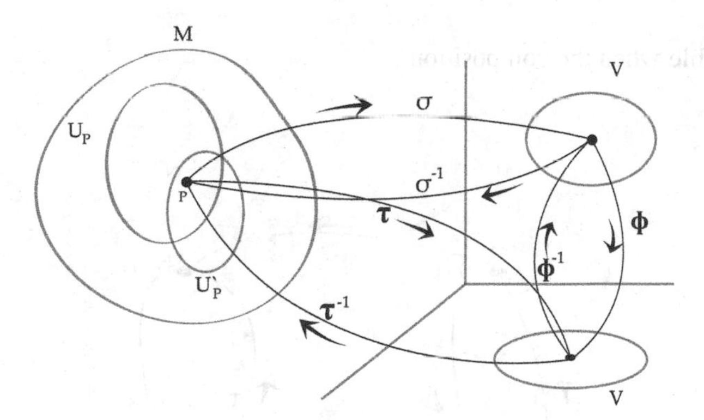

Fig. 2.1 Manifold

The maps $\sigma, \tau, \ldots$ are called *charts* and the set of all *charts* is called an atlas of $\mathcal{M}$. A *differentiable manifold* is a manifold for which ϕ is a differentiable map in $I\!R^n$. In this case we say that the differentiable manifold $\mathcal{M}$ has a differentiable atlas. The smallest n required to form an atlas is called the *dimension of the manifold.*

From the inverse σ^{-1} of each chart we may obtain a topology in $\mathscr{M}$ in the following way: Denoting by V_q an open set in $I\!R^n$, then all points in this open set are mapped by σ^{-1} in an open set U_p in $\mathscr{M}$ (Fig. 2.1). Thus, we obtain the borrowed topology in $\mathscr{M}$, where all topological properties of $I\!R^n$ are transferred to $\mathscr{M}$, including the Hausdorff property meaning that for each object in $\mathscr{M}$ there is a neighborhood containing another object of $\mathscr{M}$.

The simplest examples of manifolds are the already known curves and surfaces of $I\!R^3$. The coordinate space $I\!R^n$ itself is a trivial manifold, whose charts are identity maps. Less trivial examples are the space–times as we shall see later.

A *differentiable map* between two arbitrary manifolds can be defined through the use of the borrowed topology as follows: Let $\mathscr{M}$ and $\mathscr{N}$ be manifolds with dimensions m and n, respectively. A map $F : \mathsf{U}_p \to \mathsf{U}_q$, with $\mathsf{U}_p \in \mathscr{M}$ and $\mathsf{U}_q \in \mathscr{N}$, is said to be differentiable if for any chart σ in $\mathscr{M}$, and any chart τ in $\mathscr{N}$, the composition

$$\tau \circ F \circ \sigma^{-1} : \mathsf{V} \to \mathsf{V}'$$

is a differentiable map from $I\!R^m$ to $I\!R^n$. A *homeomorphism* F between manifolds is an invertible map such that $\tau \circ F \circ \sigma^{-1}$ is continuous. If this map is also differentiable then F is called a *diffeomorphism* (Fig. 2.2).

As an example consider that $\mathscr{M} \equiv I\!R$ and $\mathscr{N}$ is an arbitrary manifold. Then it follows from the above definition that the map

$$\alpha : \mathsf{U}_t \to \mathsf{U}', \quad \mathsf{U}' \in \mathscr{N}, \quad t \in \mathsf{U}_t \subset I\!R$$

is differentiable when the composition

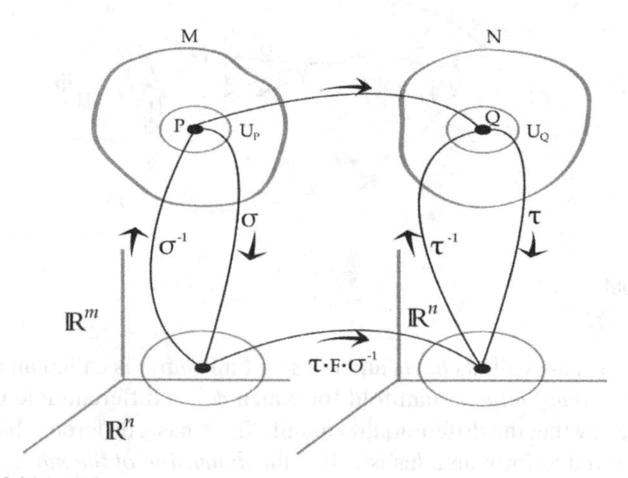

Fig. 2.2 Manifold mappings

$$I \circ \alpha \circ \sigma^{-1} = \alpha \circ \sigma^{-1} : I\!R \to I\!R^n$$

is differentiable (here the chart of $I\!R$ is the identity map I).

A *continuous curve* in $\mathscr{N}$ is a simple continuous map $\alpha(t) : I\!R \to \mathscr{N}$. A *differentiable curve* in $\mathscr{N}$ occurs when the map α is differentiable. If in addition the derivative $d\alpha/dt$ does not vanish, we have a *regular curve* in $\mathscr{N}$. From Fig. 2.3 we see that the curve in $\mathscr{N}$ is the image of a curve in $I\!R^n$ by the inverse chart. In particular, when $I\!R$ is replaced by one of the coordinate axis x^α of the $I\!R^n$, the curve $\alpha(x^\alpha)$ is called the *coordinate curve* in the manifold, whose parameter is the coordinate itself x^α.

From the definition it follows that in general a manifold is not a vector space. Therefore the notions of force, pressure, momenta, and other physical fields that depend on the specification of a direction on different points of a manifold are not defined. This may seem conflicting with the concept of a manifold as a set of observations because these observations involve interactions or forces. Vectors and vector fields are implemented in the differentiable structure of manifolds in the form of tangent vectors.

Definition 2.2 (The Tangent Bundle) A *tangent vector* to a manifold $\mathscr{M}$ at a point p is a tangent vector to a curve on $\mathscr{M}$ passing through p. To define a tangent vector to a curve on $\mathscr{M}$, consider the set of all differentiable functions defined in $\mathscr{M}$, $\mathscr{F}(\mathscr{M})$, and $f \in \mathscr{F}(\mathscr{M})$. The tangent vector field to the curve $\alpha(t)$ at the point $p = \alpha(t_0)$ can be defined by the operation

$$\frac{d}{dt} f(\alpha(t))\rfloor_{t_0} = \frac{\partial f}{\partial x^\beta} \frac{d\alpha^\beta}{dt} \rfloor_p$$

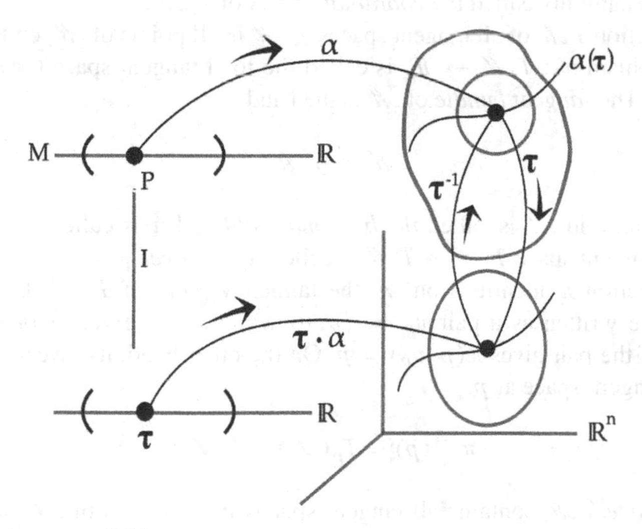

Fig. 2.3 Curve on a manifold

The derivative $d/dt\, f(\alpha(t))\rfloor_{t_0}$ is called the *directional derivative* of f with respect to the vector $\alpha'(t_0) = v_p$. It is also denoted by

$$\alpha'(t_0)[f] = v_p[f] = \frac{df}{dt}\rfloor_p$$

The set of all tangent vectors to $\mathcal{M}$ at p generates a *tangent space*, denoted by $T_p\mathcal{M}$, with respect to the vector addition rule at p: if $v_p = \alpha'(t_0)$ and $w_p = \beta'(t_0)$ are tangent vectors to two curves passing through p, then the linear combination $mv_p + nw_p = u_p$ defines another curve $\gamma(t)$ in $\mathcal{M}$ with tangent $\gamma'(t_0) = u_p$ passing through the same point $\gamma(t_0) = p$. Clearly such rule does not apply to tangent vectors in different points of $\mathcal{M}$, so that tangent vectors and tangent spaces to a manifold are only locally defined. In some textbooks a tangent vector at p is called a *vector applied to a point*.

Since $\mathcal{M}$ has dimension n, $T_p\mathcal{M}$ has dimension n and a basis of $T_p\mathcal{M}$ is composed of n linearly independent vectors, tangent to n curves in $\mathcal{M}$. In particular, these curves can be taken to be the curves defined by the coordinates x^α with tangent vectors

$$e_\alpha[f] = \alpha'(x^\alpha)[f]\rfloor_p = \frac{\partial f}{\partial x^\beta}\frac{dx^\beta}{dx^\alpha}\rfloor_p = \frac{\partial f}{\partial x^\alpha}\rfloor_p$$

Since this applies to all differentiable functions we may omit f and write the tangent basis as an operator

$$e_\alpha = \frac{\partial}{\partial x^\alpha}$$

Such basis is naturally called the *coordinate basis* of $T_p\mathcal{M}$.

The collection $T\mathcal{M}$ of all tangent spaces to $\mathcal{M}$ in all points of $\mathcal{M}$, endowed with a diffeomorphism $\pi : T\mathcal{M} \to I\!R$, is called the total tangent space (or simply the *total space*). The *tangent bundle* of $\mathcal{M}$ is the triad

$$(\mathcal{M}, \pi, T\mathcal{M})$$

where the manifold $\mathcal{M}$ is called *the base manifold* and π is called *the projection map*. Each tangent space $T_p\mathcal{M} \in T\mathcal{M}$ is called a *fiber over* p.

The projection π identifies on $\mathcal{M}$ the tangency point of $T_p\mathcal{M}$. Each tangent vector can be written as a pair $v_p = (p, v)$ while v is the vector properly. The projection of the pair gives $\pi(p, v_p) = p$. On the other hand, its inverse π^{-1} gives the whole tangent space at p:

$$\pi^{-1}(p) = T_p(\mathcal{M}) \in T\mathcal{M}$$

The *total space* $T\mathcal{M}$ contains all tangent spaces in all points of $\mathcal{M}$, so that it is composed of ordered pairs like (p, v), where $p \in \mathcal{M}$ and $v \in T_p\mathcal{M}$ (Fig. 2.4).

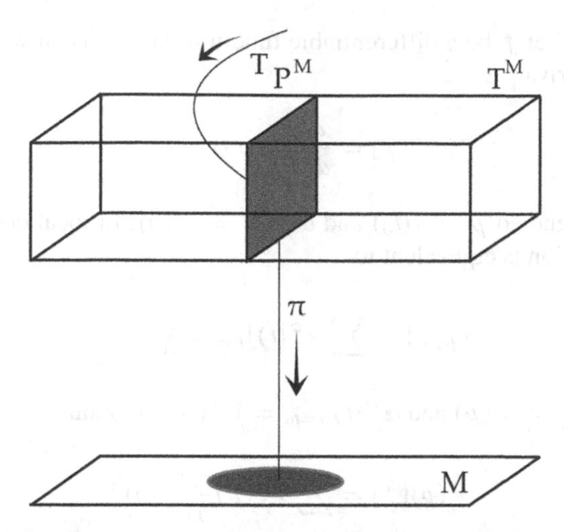

Fig. 2.4 The tangent bundle

Since $\mathcal{M}$ is a manifold with n dimensions, it follows that $T_p\mathcal{M}$ is also n-dimensional. Consequently, the set of all pairs $(p, v) \in \mathcal{M} \times T_p(\mathcal{M})$ is a manifold with dimension $2n$.

A well-known example is given by the *configuration space* of a mechanical system of idealized point particles defined in a region of a space–time $\mathcal{M}$. Supposing that all constraints to the motion are removed, we obtain a reduced representation space in which we mark ordered pairs $(x^i, \dot{x}^i)$, $i = 1..N$, where x^i denotes the coordinate of the system and $\dot{x}^i$ denotes the components of its velocity vector. This set of ordered pairs is the total space $T\mathcal{M}$ of the tangent bundle called the *representation space*.

The equations of motion of a mechanical system described in the configuration space are derived from a *Lagrangian* $\mathcal{L}(x^i, \dot{x}^i)$, which is a *differentiable function defined on the total space* $\mathcal{L} : T\mathcal{M} \to I\!R$ [55]. Classical mechanical systems evolved somewhat independently of the concept of manifold and the coordinates x^i were once called *generalized coordinates* [56].

Definition 2.3 (Tangent Vector Fields) The concept of *tangent vector field* arises naturally after the definition of the tangent bundle as a map $V : \mathcal{M} \to T\mathcal{M}$ such that it associates with each element $p \in \mathcal{M}$ a tangent vector $V(p) \in T_p\mathcal{M}$.

A *cross section* of the tangent bundle is a map $S : \mathcal{M} \to T\mathcal{M}$ such that $\pi \circ S = I$. It follows that a vector field is a particular cross section such that it specifies a vector $V(p) = v_p \in T_p(\mathcal{M})$.

Clearly, the set of vector fields on a manifold does not generate a vector space because we cannot sum vectors belonging to different tangent spaces.

The concept of *directional derivative* of a function with respect to a tangent vector can be easily extended to the directional derivative of a function with respect to a vector field: Consider a vector $v_p = V(p)$ and a curve $\alpha(t)$ such that $p = \alpha(t_0)$

and $\alpha'(t_0) = v_p$. Let f be a differentiable function on $\mathscr{M}$. Then we may calculate the directional derivative

$$v_p[f] = \frac{d}{dt} f(\alpha(t))\rfloor_{t=t_0}$$

where we have denoted $p = \alpha(t_0)$ and $\alpha'(t_0)_i = v_i(p)$. In local coordinates $\{x^\alpha\}$, the above expression is equivalent to

$$v_p[f] = \sum \alpha'^\alpha(t)\rfloor_{t=t_0} \frac{\partial f}{\partial x^\alpha}\rfloor_p$$

Thus, replacing $v_p = V(p)$ and $\alpha'^\alpha(t)|_{t=t_0} = V^\alpha(p)$ we obtain

$$V(p)[f] = \sum V^\alpha(p) \frac{\partial f}{\partial x^\alpha}(p)$$

Supposing that this holds true for all p belonging to the region of $\mathscr{M}$, we may simply suppress the point p, thus producing the directional derivative of f with respect to the vector field V in a coordinate basis:

$$V[f] = \sum V^\alpha \frac{\partial f}{\partial x^\alpha}$$

where V^α denotes the components of the vector field V in the chosen coordinates $\{x^\alpha\}$.

Consider two manifolds $\mathscr{M}$ and $\mathscr{N}$ and a differentiable map $F : \mathscr{M} \to \mathscr{N}$. The *derivative map* of F, denoted by F_*, is a *linear* map between the respective total spaces,

$$F_* : T\mathscr{M} \to T\mathscr{N}$$

such that for a differentiable function $f : \mathscr{N} \to I\!R$ and $v_p \in T_p\mathscr{M}$, the result $F_*(v_p)[f]$ is the same as the directional derivative of $f \circ F$ with respect to v_p:

$$F_*(v_p)[f] = v_p[f \circ F]$$

The linearity of F_* is a consequence of the properties of the directional derivative:

$$F_*(av_p + bw_p)[f] = (av_p + bw_p)[f \circ F] = aF_*(v_p)[f] + bF_*(w_p)[f]$$

As an example consider that $v_p = \alpha'(t_0)$ is a tangent vector to a curve $\alpha(t)$ at a point $p = \alpha(t_0) \in \mathscr{M}$. The curve $\alpha(t)$ is mapped by F to a curve of $\mathscr{N}$ given by

$$\beta(t) = F(\alpha(t))$$

By the above definition it follows that

$$F_*(\alpha'(t_0))[f] = F_*(v_p)[f] = v_p[f \circ F(\alpha(t))] = v_p[f(F(\alpha))] = v_p[f(\beta)]$$

and from the definition of the directional derivative we obtain

$$v_p[f(\beta(t))] = \frac{d}{dt}[f(\beta(t))]|_{t=t_0} = \beta'(t_0)[f]$$

so that

$$F_*(\alpha'(t_0))[f] = \beta'(t_0)[f]$$

In other words, if $\beta = F(\alpha)$, then the tangent vector to β at the point $F(p) \in \mathscr{N}$ is $\beta' = F_*(\alpha'(t_0))$.

Definition 2.4 (Vector Bundle) Quite intuitively the definition of tangent bundle can be extended to the more general notion of *vector bundle* as follows: Given a manifold $\mathscr{M}$, we may attach to each point p a *local vector space* $\mathscr{V}_p$, not necessarily tangent to a curve in $\mathscr{M}$. Then we may collect these vector spaces in a total space $\mathscr{V}$, so that we can identify the point $p \in \mathscr{M}$ where $\mathscr{V}_p$ is defined, called the *fiber over p*, defined by a projection map $\pi : \mathscr{V} \rightarrow \mathscr{M}$. This *vector bundle* is represented by the triad

$$(\mathscr{M}, \ \pi, \ \mathscr{V})$$

Clearly, the tangent bundle is a particular example of vector bundle. A less trivial example is given by the *normal bundle* where the fiber over p is a vector space N_p orthogonal to the tangent spaces $T_p\mathscr{M}$. Another example of vector bundle is given by the space of matrices attached at each point of $\mathscr{M}$.

When all fibers $\mathscr{V}_p$ of a vector bundle have the same dimension, they are all isomorphic to a single vector space Σ, called the *typical fiber*. A particularly interesting case occurs when the total space is the Cartesian product $\mathscr{V} = \mathscr{M} \times \Sigma$, the vector bundle is called a product vector bundle, or simple *product bundle*, written as

$$(\mathscr{M}, \ \pi, \ \mathscr{M} \times \Sigma)$$

In this case, the total space $\mathscr{M} \times \Sigma$ can be graphically represented by a box, which represents the fiber bundle, with $\mathscr{M}$ in the base and Σ in the vertical side. Each element of this total space is just the pair (p, v) where the vector v represents any vector in each fiber. Because of this, these vector bundles are sometimes referred to as *trivial vector bundles*.

2.2 Geometry of Manifolds

A manifold does not necessarily come with a geometry, that is, with a measure of distances or of angles, so that we may draw parallel lines satisfying Euclid's axioms. A geometry can be implemented on a manifold as follows[1]:

Definition 2.5 (Metric Geometry on a Manifold) The most intuitive way to construct parallel lines in the Euclidean space is to use a graduated rule or metric geometry. This intuitiveness is a consequence of the fact that $I\!R^3$ is a manifold and also a vector space in which a scalar product is globally defined.

To define the same notion of parallels in a manifold $\mathcal{M}$ is a little more complicated. First, we need to define the metric by the introduction of a scalar product of vectors on the manifold. Since manifolds do not have vectors, we may locally define the metric in each tangent space as a map

$$< , >: T_p\mathcal{M} \times T_p\mathcal{M} \rightarrow I\!R$$

such that it is (a) bilinear and (b) symmetric. There is a third condition in Euclidean geometry which says that it should be positive definite: Given a vector v, then (c) $||v||^2 =< v, v >\geq 0$, and $||v||^2 =< v, v >= 0 \Longleftrightarrow v = 0$. This condition is omitted when we consider that geometry is an experimental science, whose results depend on the definition of the observers, of the observed object, and of the methods of observations. Thus the condition (c) may hold under certain measurements and not in others.[2]

Since the scalar product is locally defined, the metric components in an arbitrary basis

$$g_{\mu\nu} =< e_\mu, e_\nu >$$

are also locally defined. This makes it difficult to define distances between two distinct points of the manifold connected by a curve $\alpha(t)$, for in principle the metric varies from point to point. Therefore, the comparison of distances in different points requires an additional condition that the line element

$$ds^2 = g_{\mu\nu}dx^\mu dx^\nu$$

remains the same. Such *isometry* exists naturally in Galilean, Newtonian, and Minkowski's space–times, but there is no preliminary provision for it in general relativity. In this case (as in arbitrary metric manifolds), the metric components vary

[1] A geometry can be of two basic kinds: The *metric geometries* based on the notion of distance or a graduated rule; the other is the *affine geometries* based on the notion of parallel transport of a vector field along a curve, keeping a constant angle with the tangent vector to that curve [5, 57].

[2] In mathematical analysis when the condition (c) is omitted the analysis is referred to as analysis in Lorentzian manifolds.

from point to point, *so that the measurements of distances between lines depend on the existence of an affine connection which is compatible with the metric geometry.* This affine connection was defined by Levi-Civita, using the Christoffel symbols (see below).

Definition 2.6 (Affine Geometry) An affine geometry on a manifold $\mathcal{M}$ is defined by the existence of parallel transport of a vector field W along a curve $\alpha(t)$ on $\mathcal{M}$, such that the angle between W and the tangent vector to $\alpha(t)$ remains constant. Therefore, it offers an alternative but essential way to trace parallel lines in a manifold prior to the definition of a metric. Let us detail how this works.

Given a vector field W on a manifold $\mathcal{M}$, its *covariant derivative* with respect to the vector field $V = \alpha'(t)$ tangent to a curve $\alpha(t)$ at a point $p = \alpha(t_0)$ is the measure of the variation of W along α:

$$\nabla_V W(p) = \frac{d}{dt} W(\alpha)\bigg|_{t=t_0} \tag{2.1}$$

satisfying the following properties (a and b are numbers and f is a real function defined on $\mathcal{M}$):

(a) $\nabla_V(aW + bW') = a\nabla_V W + b\nabla_V W'$
(b) $\nabla_{aV+bV'}(W) = a\nabla_V W + b\nabla_{V'} W$
(c) $\nabla_V f = V[f]$
(d) $\nabla_V(fW) = V[f]W + f\nabla_V W.$

These properties correspond to similar properties that hold in the particular case of $I\!R^n$, when we use arbitrary base vectors [58]. It is clear from the above definition that the covariant derivative of a vector field in $\mathcal{M}$ with respect to a tangent vector of $T_p\mathcal{M}$ is again a tangent vector field of the same space. It is also clear that it does not depend on the previous existence of a metric.

The above definition of covariant derivative can be easily extended to the region of definition of the involved vector fields, without specifying the point $p = \alpha(t_0)$. Denoting by $V = \alpha'(t)$ the tangent vector field to a curve $\alpha(t)$, then (2.1) gives

$$\nabla_{\alpha'} W = \frac{d}{dt} W(\alpha)$$

providing a measure of how the vector field W varies along the curve $\alpha(t)$.

Definition 2.7 (Parallel Transport) In the case when

$$\nabla_{\alpha'} W = \frac{d}{dt} W(\alpha) = 0$$

we say that the field W is *parallel transported* along $\alpha(t)$.

Thus, the existence of a *covariant derivative* is intimately associated with the existence of an affine geometry, and the covariant derivative operator ∇ is also referred to as the *affine connection operator*.

Let $\{e_\alpha\}$ be a set of n tangent vector fields to M, such that at each point p, $\{e_\alpha(p)\}$ is a basis of $T_p(M)$. Such basis is sometimes referred to as a *field basis*. Then the covariant derivative of e_α with respect to another field basis e_β is a linear combination of the same field basis:

$$\nabla_{e_\alpha} e_\beta = \Gamma^\gamma_{\alpha\beta} e_\gamma \tag{2.2}$$

where the coefficients $\Gamma^\gamma_{\alpha\beta}$ are called the connection coefficients or the *Christoffel symbols*. By different choices of the way in which the covariant derivative acts on the basis, we obtain different geometries. Thus, for example we can have Riemann, Weyl, Cartan, Einstein–Cartan, and Weitzenbock geometries, depending on the properties of these coefficients.

In the case of the Riemann geometry, the connection coefficients $\Gamma^\gamma_{\alpha\beta}$ are symmetric in the sense that

$$\nabla_{e_\alpha} e_\beta = \nabla_{e_\beta} e_\alpha$$

or equivalently, the symmetry is explicit in the two lower indices of the Christoffel symbols:

$$\Gamma^\gamma_{\alpha\beta} = \Gamma^\gamma_{\beta\alpha}$$

Here and in the following we use the choice of Riemann and Einstein, with a symmetric connection.

In order to write the *components of the covariant derivative*, let us write the vector fields in an arbitrary field basis: $W = W^\alpha e_\alpha$ and $V = V^\beta e_\beta$. From the above properties of the covariant derivatives, we obtain

$$\nabla_V W = \nabla_V (W^\alpha e_\alpha) = V[W^\alpha] e_\alpha + W^\alpha \nabla_V e_\alpha = \left(V^\beta \frac{\partial W^\gamma}{\partial x_\beta} + W^\alpha V^\beta \Gamma^\gamma_{\alpha\beta} \right) e_\gamma$$

where in the last expression we have made a convenient change in the summing indices.

Taking in particular $V = e_\alpha$, and using the semicolon to denote the components of the covariant derivative, it follows that

$$\nabla_{e_\mu} W = W^\beta{}_{;\mu} e_\beta$$

where we have denoted the components of the covariant derivative of W as

$$W^\beta{}_{;\mu} = \left(\frac{\partial W^\beta}{\partial x^\mu} + W^\gamma \Gamma^\beta_{\gamma\mu} \right) \tag{2.3}$$

The affine geometry can be made compatible with the metric geometry under the condition that the metric behaves as a constant with respect to the covariant derivative. This is what Riemann did when he postulated that the covariant derivative of the metric tensor g is zero:

$$(\nabla_{e_\rho} g)_{\mu\nu} = 0 \tag{2.4}$$

This is called the *metricity condition* of the affine connection, and it is often written in terms of the components as $g_{\mu\nu;\rho} = 0$. As we recall from the introduction, this condition was tentatively modified by Weyl in his 1919 theory.

2.3 The Riemann Curvature

The geometry of surfaces of $I\!R^3$ tells us *that the shape of a surface depends on how it deviates from the local tangent plane*. This characterizes a topological property of the surface, allowing to distinguish, for example, a plane from a cylinder. This variation of the local tangent plane can be studied alternatively by the variation of the normal vector field to the surface, and it is called the *extrinsic curvature* of the surface. It is extrinsic because it depends on a property that lies outside the surface.

Definition 2.8 (The Riemann Tensor) Consider two curves in a manifold $\mathscr{M}$, α and β intersecting at a point A, with unit *tangent independent vectors* U and V respectively. Then make a parallel displacement of V and U along the curves α and β, respectively, as indicated in Fig. 2.5. At the points B and C draw the curves α_1 and β_1 with tangent vectors parallel to U and V, respectively, obtaining the parallelogram. Next, consider a third vector field W, linearly independent from U and V, at the

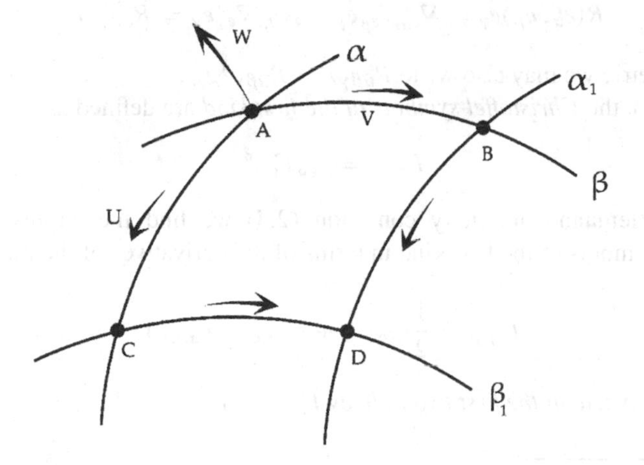

Fig. 2.5 The Riemann curvature

point A, and drag it along the curve β from A to B. Then drag it from B to D along the curve α_1. The result of such operation is the vector field

$$W' = \nabla_U \nabla_V W$$

On the other hand, dragging W from A to C and from C to D we obtain another vector

$$W'' = \nabla_V \nabla_U W$$

The difference $W' - W''$ gives the Riemann curvature tensor[3] of $\mathcal{M}$ [43, 47]

$$R(U, V)W = (\nabla_U \nabla_V - \nabla_V \nabla_U)W = [\nabla_U, \nabla_V]W \tag{2.5}$$

As we see, this result does not depend on a metric, and from our previous comment, it is actually necessary to be so before any notion of constant distance is defined.

In the particular case of a flat plane of $I\!R^3$ the Riemann tensor vanishes. Therefore, Riemann's idea of curvature is compatible with the geometry of surfaces in $I\!R^3$, at least for some basic figures. *However, it is not sufficient to distinguish a plane from a cylinder or, in fact, from an infinite variety of ruled surfaces.* It is also interesting to note that for surfaces of $I\!R^3$ the Riemann tensor coincides with the *Gaussian curvature* $K = k_1 k_2$, where k_1 and k_2 are the *principal curvatures* measured by the maximum and minimum deviations of the normal vector field (see, e.g., [48, 58]). The Egregium theorem of Gauss shows that indeed K can be defined entirely as an intrinsic property of the surface.

The components of the Riemann tensor of a manifold $\mathcal{M}$ in an arbitrary tangent basis $\{e_\mu\}$ can be obtained from (2.5) when the operator is applied to the basis vectors, reproducing another vector

$$R(e_\alpha, e_\beta)e_\gamma = \nabla_{e_\alpha} \nabla_{e_\beta} e_\gamma - \nabla_{e_\beta} \nabla_{e_\alpha} e_\gamma = R_{\alpha\beta\gamma}{}^{\delta} e_\delta \tag{2.6}$$

Using the metric we may also write $R_{\alpha\beta\gamma\delta} = R_{\alpha\beta\gamma}{}^{\delta} g_{\delta\varepsilon}$.

From (2.2), the *Christoffel symbols of the first kind* are defined as

$$\Gamma_{\alpha\beta\gamma} = g_{\gamma\delta} \Gamma_{\alpha\beta}{}^{\delta}$$

and using Riemann's metricity condition (2.4) we find the expression of the Christoffel symbols of the first kind in terms of the derivatives of the metric

$$\Gamma_{\alpha\beta\gamma} = \frac{1}{2}(g_{\alpha\gamma,\beta} + g_{\beta\gamma,\alpha} - g_{\alpha\beta,\gamma})$$

which is *symmetric in the first two indices* $\Gamma_{\alpha\beta\gamma} = \Gamma_{\beta\alpha\gamma}$.

---

[3] In general U, V, W need not be linearly independent, but in this case we need to add the term $\nabla_{[U,V]} W$ to compensate for the linear dependency in the construction of the parallelogram.

Replacing these components in (2.5) we obtain the components of the Riemann tensor:

$$R_{\alpha\beta\gamma\varepsilon} = \Gamma_{\beta\varepsilon\alpha;\gamma} - \Gamma_{\beta\varepsilon\gamma;\alpha} + \Gamma^{\mu}_{\beta\gamma}\Gamma_{\alpha\varepsilon\mu} - \Gamma^{\mu}_{\beta\alpha}\Gamma_{\gamma\varepsilon\mu} \tag{2.7}$$

From this expression we derive the following properties:

$$R_{\alpha\beta\gamma\varepsilon} = -R_{\beta\alpha\gamma\varepsilon} \tag{2.8}$$

$$R_{\alpha\beta\gamma\varepsilon} = -R_{\alpha\beta\varepsilon\gamma} \tag{2.9}$$

$$R_{\alpha\beta\gamma\varepsilon} = R_{\gamma\varepsilon\alpha\beta} \tag{2.10}$$

$$R_{\alpha\beta\gamma\varepsilon} + R_{\alpha\varepsilon\beta\gamma} + R_{\alpha\gamma\varepsilon\beta} = 0 \tag{2.11}$$

Finally the covariant derivative of Riemann's tensor gives the Bianchi's identities

$$R_{\alpha\beta\gamma\varepsilon;\mu} + R_{\alpha\beta\varepsilon\mu;\gamma} + R_{\alpha\beta\mu\gamma;\varepsilon} = 0 \tag{2.12}$$

Ricci's curvature tensor is derived from Riemann's tensor by a contraction

$$R_{\alpha\varepsilon} = g^{\beta\gamma} R_{\alpha\beta\gamma\varepsilon} \tag{2.13}$$

On the other hand, the contraction of Ricci's tensor gives the scalar curvature (or the Ricci scalar curvature).

$$R = g^{\alpha\beta} R_{\alpha\beta} \tag{2.14}$$

We shall return to the Riemann tensor in the latter sections, showing that it has the same structure for gravitation and for the gauge field strengths.

Example 2.1 (Geodesic) A *geodesic* in a manifold $\mathscr{M}$ is a curve such that its tangent vector is transported *parallel to itself*:

$$\nabla_{\alpha'}\alpha' = 0$$

From (2.3) we may derive the equation of a geodesic $\alpha(t)$, with parameter t, in coordinate basis. Taking $V = W = \alpha' = \sum x^{\mu} e_{\mu}$, and using the geodesic definition, we obtain

$$\frac{d^2x^{\mu}}{dt^2} + \Gamma^{\mu}_{\alpha\beta}\frac{dx^{\alpha}}{dt}\frac{dx^{\beta}}{dt} = 0 \tag{2.15}$$

In particular, for $\mathscr{M} = I\!R^n$ this is the equation for a straight line in arbitrary coordinates.

As an exercise on the equivalence between metric and affine geometries under (2.4), let us show that geodesics generalize the concept of straight lines in the sense that describe the smallest distance between two points of $\mathscr{M}$, measured by a metric associated with a scalar product $<, >$.

Consider a family of curves passing through two arbitrary points p and q in $\mathscr{M}$, defined by the displacement of a vector field W over the geodesic α:

$$\gamma(t, u) = \alpha(t) + uW(\alpha(t))$$

It follows that $\gamma' = d\gamma/dt = \alpha' + udW/dt$ and $d\gamma/du = W$. The arc-length between p and q along any curve of the family is given by

$$S(u) = \int_o^t \sqrt{< \gamma', \gamma' >} dt$$

The variation of this arc-length with respect to the family parameter u is

$$\frac{dS}{du} = \int_0^t \frac{< \frac{d\gamma'}{du}, \gamma' >}{\sqrt{< \gamma', \gamma' >}} dt$$

Since W is an arbitrary vector field we may take in particular $W = \alpha'$, so that $d\gamma/du = W = \alpha'$. Using the fact that the two parameters are independent we obtain

$$\frac{d\gamma'}{du} = \frac{d}{dt}\frac{d\gamma}{du} = \frac{d\alpha'}{dt} = \nabla_{\alpha'}\alpha'$$

Since α is a geodesic, we necessarily have

$$\frac{dS}{du} = \int_0^t \frac{< \nabla_{\alpha'}\alpha', \gamma' >}{\sqrt{< \gamma', \gamma' >}} dt = 0$$

showing that S is a maximum or a minimum. The maximum is infinity and therefore it is not interesting. The minimum occurs in the geodesic.

Chapter 3
Symmetry

Weyl's classic book on symmetry conveys the idea that the notion of symmetry is not just an art or an invention of the mind, but part of the observational structure of nature [59]. However, the awareness of the importance of symmetry in physics became clear only after the debate on the negative result of the Michelson–Morley experiment and the subsequent interpretation of the relative motion between observers and observables given by Einstein. This interpretation led us to the emergence of the Poincaré symmetry. From then on, the structure of *Lie symmetry* has become the essential tool to the understanding of the fundamental interactions.

3.1 Groups and Subgroups

A group G is a set composed of elements $a, b, c, \ldots$, endowed with a closed operation (generically denoted by $*$) such that

(a) The operation is associative: $a * (b * c) = (a * b) * c$;
(b) There is a neutral or identity element 1: $a * 1 = 1 * a = a$; and
(c) For each element $a \in G$, there is an inverse element denoted by a^{-1} such that $a^{-1} * a = a * a^{-1} = 1$.

For an Abelian or commutative group we also have $a * b = b * a$. The number of elements in a group is called the *order of the group*. A group of infinite order has infinite elements.

A subset $H \subset G$ is a *subgroup* of G, when its elements form a group *with the same operation of* G. If $H \subset G$, $H \neq G$, then H is a *proper subgroup* of G. It is easy to see that the identity of G must be also contained in all subgroups of G.

Like a manifold, a group can be parameterized by a set of real numbers $(\theta^1, \ldots, \theta^N)$, given by 1:1 maps or charts $X : G \to I\!R^N$ such that for each element $r \in G$ we have an element of $I\!R^N$

$$X(r) = (\theta^1, \ldots, \theta^N)$$

and conversely, given a point in $I\!R^N$ we obtain an element of G

M.D. Maia, *Geometry of the Fundamental Interactions*,
DOI 10.1007/978-1-4419-8273-5_3, © Springer Science+Business Media, LLC 2011

$$r = X^{-1}(\theta^1, \ldots, \theta^N) = X^{-1}(\theta)$$

The *dimension of a group* is the maximum number N of independent parameters required to describe any element of the group.

Given two elements $r = X_1(\theta)$ and $s = Y^{-1}(\theta')$, the group composition gives $r * s = X^{-1}(\theta) * Y^{-1}(\theta') = t = Z^{-1}(\theta'') \in G$. The parameters θ'' must then be related to θ and θ' as

$$\theta'' = Z \circ (X^{-1}(\theta) * Y^{-1}(\theta')) = f(\theta, \theta') \tag{3.1}$$

Since these charts cover the whole group, they form an *atlas* similar to the case of differentiable manifolds. Then the above condition (3.1) must be satisfied for all elements of the group (such condition does not exist in differentiable manifolds).

When the parameters vary continuously within a given interval on $I\!R^N$ and (3.1) is a homeomorphism we have a *continuous group*. This is less demanding than the differentiable manifold structure where the relation between the parameters (the coordinates) is a diffeomorphism. However, later on we shall be using an even stronger condition imposed by Lie, where (3.1) is required to be an analytic function.

For notational simplicity, from now on we will omit the $*$ operation and write it simply as a product.[1] Thus $r * s$ is written simply as rs.

Definition 3.1 (Cosets and Normal Subgroups) Consider a subgroup of $H \subset G$ and r a specific element of G. Then the set denoted by

$$rH = \{rx \,|\, x \in H\}$$

is called the *left coset* of H. Similarly we may define the *right coset* of H denoted by Hr. It follows from this definition that $rH \neq H$ because r is not necessarily in H. However, rH necessarily contains r because as a subgroup H contains the identity 1. Hence $r = r1 \in rH$.

Given two left cosets (or right cosets) of a subgroup H in G, aH and bH, if they possess a common element then they are necessarily identical.

Indeed, consider the left cosets $A = aN$ and $b = bN$ and let x be a common element belonging to A and B. Then we may write $x = ar$, $x = bs$, $r, s \in H$, it follows that $ar = bs$ and therefore, $a = bsr^{-1} = bt$, $t = sr^{-1} \in H$. Consequently,

$$aH = \{am \,|\, m \in H\} = \{btm \,|\, m, t \in H\} = \{bn \,|\, n \in H\} = bH$$

A subgroup N in G such that its left and right cosets are equal is called an *invariant* (or *normal*) subgroup of G. From the defining condition $aN = Na$, it follows that

[1] Except for additive groups where the operation is a sum and the neutral element is zero.

$$N = a^{-1}Na = \{q = a^{-1}xa \,|\, x \in N,\ a \in G\}$$

In other words, the elements of an invariant subgroup belong to a class of equivalence where they differ by an equivalence relationship $q \sim x$ defined by $q = a^{-1}xa$.

An interesting property is that if $A = aN$ and $B = bN$ are cosets of an invariant subgroup N, then the set $C = AB = \{xy \in G \,|\, x \in A \text{ and } y \in B\}$ is also a coset of G. Indeed, writing $x = ap$ and $y = bq,\ p, q \in N$, it follows that the elements of C have the form $xy = apbq$. Since N is an invariant subgroup the left and right cosets of N are identical. Hence, if $p, q \in N$ then $aga^{-1} = r$ and $bab^{-1} = a$ where $r, s \in s$. Therefore, $ap = cp$ and $bs = sb$ and $af = apbq = apsb$. However, $p, s \in N,\ ps \in N$, and using again the fact that N is invariant, $psb = bm,\ m \in N$. Consequently, $xy = abm = cm,\ c = ab$, which implies that xy belongs to a left coset of N, $C = AB = cN$.

The above result suggests the construction of a product operation between cosets of a group G as follows: Given two left cosets A and B defined by the same invariant subgroup N in G, then $C = AB$ is also a left coset cN where $c = ab,\ A = aN$ and $B = bN$.

It can be easily seen that this product defines a group, where the identity element is N:

$$A = AN = \{xy = cm \,|\, c = a1,\ m \in N\} = \{z = am \,|\, m \in\} = A$$

the inverse of $A = aN$ is $A^{-1} \stackrel{\text{def}}{=} a^{-1}N$:

$$C = AA^{-1} = \{xy = cm|c = aa^{-1} = 1,\ m \in N\} = \{z = m|m \in N\} = N.$$

Finally, the product of cosets is associative: If $A = aN,\ B = bN,\ C = cN$, then $(AB)C = (ab)cN = a(bc)N = A(BC)$.

The set of all cosets of an invariant subgroup N, like $A = aN$, defines a group, called the quotient group G/N, with respect to the above defined coset product.

3.2 Groups of Transformations

Groups can be studied by themselves as abstract groups. On the other hand, symmetry groups are *transformation groups*, whose elements are *operators acting on a space or manifold.* We shall be dealing mostly with transformation groups, separated in two cases which are of immediate interest to field theory and to the fundamental interactions. They are *groups of coordinate transformations* acting on the coordinates of a space–time manifold and the *groups of field transformations* acting on field variables.

The groups of *coordinate transformations* act on the coordinate spaces of the manifolds, changing a given coordinate system to another as

$$x'^i = f^i(x^1, x^2, \ldots, x^n, \theta^1, \theta^2, \ldots, \theta^N) = f^i(x^\mu, \theta^a) \tag{3.2}$$

where in the abbreviated notation x^μ are the old coordinates and θ^a are the parameters of the group.

A simple example of a coordinate transformation group is given by a group of *linear operators* acting on the parameter space $I\!R^3$ of some three-dimensional manifold. If (e_μ) is an arbitrary basis of $I\!R^3$, then the action of the group on that space can be obtained by the action of the group on that basis. For a linear operator $r \in G$ the result is a linear combination of the same basis elements:

$$r(e_i) = r_i^j e_j$$

The quantities r_i^j define a matrix in that basis which represents the group action.

Definition 3.2 (Representations of a Group) A representation of an abstract group G by a transformation group G' is a *homomorphism* $\mathscr{R} : G \to G'$. In the product notation the homomorphism writes as

$$\mathscr{R}(xy) = \mathscr{R}(x)\mathscr{R}(y) \tag{3.3}$$

Of particular importance is a *linear representation of a group* G, which is the homomorphism

$$\mathscr{R} : G \to G'$$

where G' is a *group of linear operators* acting on some vector space V, called the *representation space*.

Therefore, for each element r of the group G there is a corresponding operator $\mathscr{R}(r)$ acting linearly on the *representation space*. Since $\mathscr{R}$ is a homomorphism,

$$\mathscr{R}(rs) = \mathscr{R}(r)\mathscr{R}(s)$$

From the properties of groups it follows that $\mathscr{R}(r) = \mathscr{R}(r1) = \mathscr{R}(r)\mathscr{R}(1)$. Hence $\mathscr{R}(1) = 1$ and $\mathscr{R}(r^{-1}) = \mathscr{R}(r)^{-1}$.

Therefore, to find a linear representation of a given group, we need in the first place to define a representation space where a transformation group G' acts linearly. Then determine how the group G' acts on that space. Finally, choose a basis of the representation space.

Generally speaking, given one basis $\{\eta_i\}$ in the representation space, the linear representation is defined by the coefficients $\mathscr{R}^i_j$ in the operation

$$\mathscr{R}(r)\eta_i = \mathscr{R}_i^j(r)\,\eta_j$$

Note also that a homomorphism between groups is not necessarily a 1:1 map. In particular we may have several objects in G' corresponding to the identity element of G. The set of all such elements is called the *kernel of the representation*. Denoting this kernel by K, the above definition says that $\mathscr{R}(K) = 1$.

A *faithful linear representation* of a group is a linear representation which is 1:1. That is,

$$\mathscr{R}(r) = \mathscr{R}(s) \iff r = s$$

In this case the kernel contains only the identity element: $K = \{1\}$.

3.3 Lie Groups

A *continuous group* is such that its elements vary continuously with its parameters. The relation between parameters (3.1) can be just a homeomorphism: continuous with an inverse which is also continuous.

Like in a manifold a continuous group may have an induced topology from its parameter space, so that the operations of limits and derivatives can be defined. From this topology it is easy to infer that we may define continuous curves on a continuous group G as a continuous map $\alpha : I\!R \rightarrow G$, with tangent vector at a point $r = \alpha(t_0) \in G$, given by $\alpha'(t) = d\alpha/dt\rfloor_{t_0}$ as long as the relation between the parameters (3.1) remains valid.

Consequently we may define on a continuous group some topological properties and classify them according to topological characteristics such as

1. When any two elements of a continuous group G can be connected by any continuous curve or by a continuous sequence of segments of continuous curves, then G is called a *connected group*.
2. A group G is *multiple connected* when there are multiple curves connecting any two elements of G, but they cannot be continuously deformed into one another.
3. A group is compact when each of its parameters θ^a varies in a closed and limited interval.

Definition 3.3 (Lie Group) A continuous group G is a *Lie group* when the composition between the parameters (3.1) is analytic in the sense that $f(\theta, \theta')$ can be represented by converging positive power series [60]. We will see the relevance of this condition when discussing Lie's theorem.

A coordinate transformation produced by a Lie group acting on a differentiable manifold can be written as

$$x'^{\mu} = f^{\mu}(x^{\nu}, \theta^a) \tag{3.4}$$

where $f^{\mu} = x'^{\mu}$ are differentiable functions of the coordinates x^{μ}, but as a consequence of the analyticity of (3.1), they are analytic functions of the parameters θ^a. The local inverse transformation can be either postulated or derived from the condition that the transformation (3.4) is also *regular* in x, that is, when the Jacobian matrix

$$J(f) = \left(\frac{\partial x'^{\mu}}{\partial x^{\nu}}\right)$$

is non-singular.

A natural linear representations of the Lie group of a coordinate transformations on a manifold is given by the action of the group in the tangent and cotangent spaces, using, respectively, the coordinate basis $\{e_\mu = \partial/\partial x^\mu\}$ and its dual $\{e^\mu = dx^\mu\}$. Taking an arbitrary element $r \in G$, its action on the coordinates also changes the tangent basis as

$$e'_\mu = \frac{\partial}{\partial x'^{\mu}} = \mathscr{R}(r)e_\mu = \sum \mathscr{R}(r)_\mu{}^\nu e_\nu = \sum \mathscr{R}(r)_\mu{}^\nu \frac{\partial}{\partial x^\nu}$$

where $\mathscr{R}(r)_\mu{}^\nu$ denote the matrix elements of the linear representation defined in the tangent space.

Similarly, we obtain the dual representation of the same group using the cotangent space, with the dual coordinate basis $\{e^\mu = dx^\mu\}$. In this basis the linear representation is given by

$$e'^{\mu} = dx'^{\mu} = \sum \mathscr{R}^*(r)^\mu{}_\nu e^\nu = \sum \mathscr{R}^*(r)^\mu{}_\nu dx^\nu$$

where $\mathscr{R}^*(r)_\mu{}^\nu$ denote the matrix elements of the dual linear representation defined in the cotangent space.

More generally we may consider the group G acting on the fibers $\mathscr{V}_p$ defined on an arbitrary vector bundle $(\mathscr{M}, \pi, \mathscr{V})$, where each fiber $\mathscr{V}_p$ is a vector space in which a generic field Ψ, is defined.

Like in the tangent spaces, the action of the group on these fields can be defined by its action on a field basis of these spaces. For example, denoting a basis in one such space by $\{\eta_i\}$, then the group action on a vector $\mathscr{V}_p$ can be determined by its action on that basis as

$$\eta'_i = \mathscr{R}(r)e_i = \sum \mathscr{R}(r)_i{}^j \eta_j \tag{3.5}$$

and a similar linear representation can be defined in the dual basis of the dual fiber $\mathscr{V}_p^*$.

The following table shows some examples of Lie symmetry groups which are relevant for the current development of the theory of the fundamental interactions. Some of these groups will be also discussed in the next sections. For more on exceptional groups see, e.g., [61].

Group	Name	Group elements	Parameters
	Galilean group	3 rotations + 3 boosts + 3 translations + 1 time scale	10
	General Galilean group	3 Rotations + 3 general boosts + 1 time scale + Newton's potential gauge	10
P_4	Poincaré group	6 Pseudo-rotations + 4 translations	10
C_0	Conformal group	Poincaré subgroup + SCT[a] + dilatations + inv.	15
dS_n	deSitter group	Pseudo-rotations on n-dimensional positive sphere	$n(n+1)/2$
AdS_n	Anti-deSitter groups	Pseudo-rotations on n-dimensional negative sphere	$n(n+1)/2$
$GL(N, \mathbb{R})$	Real linear group	Real $N \times N$ matrices	N^2
$SL(N)$	Special linear group	Complex $N \times N$ matrices with determinant 1	$2(N^2-1)$
$SL(N)$	Unimodular group	Real $N \times N$ matrices	N^2-1
$U(N)$	Unitary group	Unitary matrices	N^2
$SU(N)$	Special unitary group	Unitary matrices with determinant 1	N^2-1
$SO(N)$	Special orthogonal group	Real orthogonal matrices with determinant 1	$N(N-1)/2$
G_2	Smallest exceptional group	Automorphisms of octonions	14
E_8	Largest exceptional group	The symmetry group of its Lie algebra	248

[a]Special conformal transformations

3.4 Lie Algebras

The relevance of continuous groups for the study of symmetries is that they allow us to consider infinitesimal transformations defined by when the parameters are small in the presence of unity. As before, we start with the simpler case of a group of coordinate transformations on a manifold $\mathscr{M}$.

3.4.1 Infinitesimal Coordinate Transformations

Consider a coordinate transformation described in (3.4) $x'^{\mu} = f^{\mu}(x^{\nu}, \theta^a)$, followed by a second transformation to another set of coordinates close to x'^{μ}. By the continuity of the group, the parameters of this second transformation must correspond to a small deviation from θ. That is,

$$x''^{\mu} = f^{\mu}(x^{\nu}, \theta^a + \delta\theta^a)$$

Next, expand this function in a Taylor series around $\delta\theta^{\mu} = 0$. Keeping only the first power of $\delta\theta$, we obtain

$$x''^{\mu} = f^{\mu}(x^{\mu}, \theta^{a}) + \sum \frac{\partial f^{\mu}(x, \theta^{a} + \delta\theta^{a})}{\partial \theta^{a}}\bigg|_{\delta\theta=0} \delta\theta^{a} = x'^{\mu} + a_{a}^{\mu}(x)\delta\theta^{a} = x'^{\mu} + \xi^{\mu}$$

where we have denoted $\xi^{\mu} = a_{a}^{\mu}(x')\delta\theta^{a}(x, \theta)$, called the *infinitesimal descriptor* of the transformation. As a consequence of (3.1), these functions are also analytic in θ^{a}.

The array a_{a}^{μ} depend on x and θ so that the inverse transformation exists only if it has rank equal to the smallest value between N and n. Simplifying, we may drop the excess primes to write the above infinitesimal coordinate transformation as

$$x'^{\mu} = x^{\mu} + \xi^{\mu} \tag{3.6}$$

To obtain the transformations of fields consider first the infinitesimal coordinate transformation on a differentiable real function F on the manifold

$$dF = \frac{\partial F}{\partial x^{\mu}} dx^{\mu} = \frac{\partial F}{\partial x^{\mu}} a_{a}^{\mu}(x)\delta\theta^{a} = \delta\theta^{a} X_{a} F$$

where we have introduced a linear operator acting on the space of all such differentiable functions on $\mathcal{M}$ by

$$X_{a} = \sum a_{a}^{\mu}(x) \frac{\partial}{\partial x^{\mu}} \tag{3.7}$$

Using these operators the infinitesimal variation of the function can be expressed as

$$F' = F + dF = (1 + \sum \delta\theta^{a} X_{a})F$$

In particular, taking F to be any coordinate x^{μ} we obtain our previous infinitesimal coordinate transformation (3.6).

The linear operators (3.7) generate an N-dimensional vector space with the operations of sum and multiplication by numbers given by

$$(aX_{a} + bY_{a})f = aX_{a}f + bY_{a}f, \ a, b \in I\!R$$

Indeed, suppose that there are constants $c_{a} \in I\!R$, such that $\sum c_{a} X_{a} = 0$. Applying this to x^{μ}, we obtain

$$\sum c_{a} X_{a} x^{\mu} = 0$$

Replacing definition (3.7), we obtain $\sum c_{a} a_{a}^{\mu} = 0$. Since the matrix $a_{a}^{i}(x)$ has rank equal to the smallest value between N and n, it follows that $c_{a} = 0$. Therefore, *the operators X_{a} are independent and generate a vector space called the* space of the linear operators of the Lie Group G, *denoted by $\mathcal{G}$*.

3.4.2 Infinitesimal Transformations on Vector Bundles

The situation here is similar to the previous case, with the difference that G acts on the fibers $\mathscr{V}$ of an arbitrary vector bundle, not necessarily resulting from a coordinate transformation.

Denoting a generic field defined in that vector bundle by $\Psi : \mathscr{M} \to T\mathscr{V}$ and denoting a field basis η_i, we may express the field as

$$\Psi = \Psi^i \eta_i$$

The infinitesimal transformation of the field is obtained as in the case of coordinates, where instead of a transformation of the coordinates x^μ we have a transformation of the components Ψ^i by the action of the group G, denoted by

$$\Psi'^i = f^i(\Psi^j, \theta^a)$$

which is followed by another transformation close to the first, given by

$$\Psi''^i = f^i(\Psi'^j, \theta^b + \delta\theta^b)$$

Expanding f^i in Taylor series around $\delta\theta^b = 0$ and keeping only the first powers of $\delta\theta^b$ we obtain as before

$$\Psi''^i = f^i(\Psi'^j, \theta^b) + \sum \frac{\partial f^i(\Psi'^j, \theta^b + \delta\theta^b)}{\partial\theta^c}\bigg]_{\delta\theta=0} \delta\theta^c = \Psi'^i + a^i_b(\Psi)\delta\theta^b \tag{3.8}$$

where we have denoted $a^i_b(\Psi) = \frac{\partial f^i}{\partial\theta^b}\Big]_{\delta\theta=0}$. Therefore, the infinitesimal variations of the field components are

$$\delta\Psi^i = a^i_b(\Psi)\delta\theta^b \tag{3.9}$$

and the infinitesimal variation of a function (or better, of a functional of the field such as, for example, the Lagrangian), of the field $F(\Psi)$, resulting from the above infinitesimal transformation is

$$\delta F = \frac{\partial F}{\partial\Psi^i}\delta\Psi^i = \frac{\partial F}{\partial\Psi^i}a^i_b(\Psi)\delta\theta^b = \delta\theta^b X_b F$$

where we have denoted the linear operators

$$X_a = \sum a^i_a(\Psi)\frac{\partial}{\partial\Psi^i} \tag{3.10}$$

These operators act on the space of all differentiable functions on $\mathscr{M}$.

In particular, applying X_a to the field components Ψ^α we obtain

$$X_a \Psi^i = a_b^i \tag{3.11}$$

The commutator or *Lie bracket* between these operators is defined by

$$[X_a, X_b]F = X_a(X_b F) - X_b(X_a F) \tag{3.12}$$

where F is an arbitrary function. Replacing the above expressions for X_a we find that

$$[X_a, X_b] = \left(a_a^k \frac{\partial a_b^j}{\partial \Psi^k} - a_b^k \frac{\partial a_a^j}{\partial \Psi^k} \right) \frac{\partial}{\partial \Psi^j} \tag{3.13}$$

A non-trivial result obtained by Marius Sophus Lie in 1872 shows that for a Lie group the commutators between two linear operators define an algebra in the space $\mathscr{G}$ generated by $\{X_\alpha\}$. The mathematical and physical implications of Lie's theorem reside in the fact that under the conditions defining the Lie group, it is sufficient to work with the above-mentioned algebra of linear operators.

Except for a few discrete symmetries, all relevant symmetries of the fundamental interactions satisfy these conditions. *The result of Lie is part of the development of field theory and particle physics from the 20th century onward. Actually, the results derived by Noether and Wigner suggest that any present or future theoretical proposals to modify the Lie symmetry structure must be checked against the theoretical and experimental results that are currently dependent on the Lie theorem.* This classic and non-trivial theorem can be shown in different ways. In the following we present some of its details [62–64].

Theorem 3.1 (Lie) *The commutator between two linear operators X_a of a Lie group is a linear combination of these operators*

$$[X_a, X_b] = f_{ab}{}^c X_c$$

where $f_{ab}{}^c$ are constants, called the structure constants of the group.

Consider a Lie group G with parameters θ, acting on the fiber $\mathscr{V}_p$ of a vector bundle

$$(\mathscr{M}, \pi, \mathscr{V})$$

Consider two consecutive infinitesimal transformations of the Lie group, one with increment $\theta + \delta\theta$ to the original values θ and the other with increments $\theta + d\theta$. From (3.1) and the definition of a Lie group, it follows that there is an analytic function between these parameters:

$$\theta^a + d\theta^a = \phi^a(\theta, \theta + \delta\theta)$$

The first-order terms of the Taylor expansion of this function around the identity (here represented by $\theta = 0$) gives a relation between the increments $d\theta$ and $\delta\theta$ as

$$d\theta^a = \phi(0,0) + \sum \frac{\partial \phi^a(\theta, \theta + \delta\theta)}{\partial \delta\theta^b}\rfloor_{\theta=0}\, \delta\theta^b = \sum_b \mathscr{F}_b{}^a \delta\theta^b \tag{3.14}$$

where we have denoted

$$\mathscr{F}_b{}^a(\delta\theta) = \frac{\partial \phi^a(\theta, \theta + \delta\theta)}{\partial \delta\theta^b}\bigg\rfloor_{\theta=0}$$

We note that $\phi(0,0)$ is the relation between the parameter of the identity transformation ($\theta = 0$) and itself, so that $\phi(0,0) = 0$.

From the existence of the inverse element of a group, it follows that the relation (3.1) must also be invertible. That is, there exists the inverse matrix $\mathscr{F}^{-1b}{}_a(\theta)$ and the inverse relation is given by $\delta\theta^a = \mathscr{F}^{-1a}{}_b(\delta\theta) d\theta^b$.

Replacing this in the differential $d\Psi = \Psi'' - \Psi'$ given by (3.8) we obtain

$$d\Psi^i = \frac{\partial f^i(\Psi^j, \theta^b)}{\partial \theta^b} d\theta^b = \sum a^i_a \mathscr{F}^{-1a}{}_b(\theta) d\theta^b$$

from which we obtain

$$\frac{\partial f^i}{\partial \theta^b} = \sum a^i_a \mathscr{F}^{-1a}{}_b(\theta) \tag{3.15}$$

Now, since the transformation function f^i are analytic in θ they satisfy the Leibniz derivative rule, and the last expression gives

$$\frac{\partial a^i_c \mathscr{F}^{-1c}{}_b}{\partial \theta^a} = \frac{\partial a^i_c \mathscr{F}^{-1c}{}_a}{\partial \theta^b}$$

or equivalently

$$a^i_c \left(\frac{\partial \mathscr{F}^{-1c}{}_b}{\partial \theta^a} - \frac{\partial \mathscr{F}^{-1c}{}_a}{\partial \theta^b} \right) + \mathscr{F}^{-1c}{}_b \frac{\partial a^i_c}{\partial \theta^a} - \mathscr{F}^{-1c}{}_a \frac{\partial a^i_c}{\partial \theta^b} = 0 \tag{3.16}$$

However, from (3.8), a^i_c depends on θ^a only through Ψ^i, so that

$$\frac{\partial a^i_c}{\partial \theta^a} = \frac{\partial a^i_c}{\partial \Psi^j}\frac{\partial \Psi^j}{\partial \theta^a} = \frac{\partial a^i_c}{\partial \Psi^j} a^j_d \mathscr{F}^{-1d}{}_a$$

Consequently, (3.16) becomes

$$a_c^i\left(\frac{\partial \mathscr{F}^{-1c}{}_b}{\partial \theta^a}-\frac{\partial \mathscr{F}^{-1c}{}_a}{\partial \theta^b}\right)+\frac{\partial a_c^i}{\partial \Psi^j}a_d^j\mathscr{F}^{-1d}{}_a\mathscr{F}^{-1c}{}_b-\frac{\partial a_c^i}{\partial \Psi^j}a_d^j\mathscr{F}^{-1d}{}_b\mathscr{F}^{-1c}{}_a=0$$

Therefore, multiplication of this by $\mathscr{F}^a_m\mathscr{F}^b_n$ gives

$$\left(\frac{\partial \mathscr{F}^{-1c}{}_b}{\partial \theta^a}-\frac{\partial \mathscr{F}^{-1c}{}_a}{\partial \theta^b}\right)\mathscr{F}^a_m\mathscr{F}^b_n a_c^i=-a_d^j\frac{\partial a_c^i}{\partial \Psi^j}\left(\delta^d_m\delta^c_n-\delta^d_n\delta^c_m\right)=a_m^j\frac{\partial a_n^i}{\partial \Psi^j}-a_n^j\frac{\partial a_m^i}{\partial \Psi^j} \tag{3.17}$$

defining a function of the parameters θ by

$$\left(\frac{\partial \mathscr{F}^{-1c}{}_b}{\partial \theta^a}-\frac{\partial \mathscr{F}^{-1c}{}_a}{\partial \theta^b}\right)\mathscr{F}^a_m\mathscr{F}^b_n=f^c_{mn}(\theta) \tag{3.18}$$

or equivalently

$$\frac{\partial \mathscr{F}^{-1b}{}_a}{\partial \theta^c}-\frac{\partial \mathscr{F}^{-1b}{}_c}{\partial \theta^a}=f^b_{mn}(\theta)\mathscr{F}^{-1m}{}_c\mathscr{F}^{-1n}{}_a \tag{3.19}$$

(3.17) can be written in the more compact form

$$a_m^j\frac{\partial a_n^i}{\partial \Psi^j}-a_n^j\frac{\partial a_m^i}{\partial \Psi^j}=f^c_{mn}(\theta)a_c^i \tag{3.20}$$

To end the theorem we have to show that $f^c_{mn}(\theta)$ are constants. For this purpose, take the partial derivatives of the above expression with respect to θ^b, obtaining

$$\frac{\partial f^c_{mn}(\theta)}{\partial \theta^b}a_c^i+f^c_{mn}(\theta)\frac{\partial a_c^i}{\partial \Psi^k}\frac{\partial \Psi^k}{\partial \theta^b}=\frac{\partial}{\partial \Psi^k}\left(a_m^j\frac{\partial a_n^i}{\partial \Psi^j}-a_n^j\frac{\partial a_m^i}{\partial \Psi^j}\right)\frac{\partial \Psi^k}{\partial \theta^b}$$

or

$$a_c^i\frac{\partial f^c_{mn}(\theta)}{\partial \theta^b}=\left[\frac{\partial}{\partial \Psi^k}\left(a_m^j\frac{\partial a_n^i}{\partial \Psi^j}-a_n^j\frac{\partial a_m^i}{\partial \Psi^j}\right)-f^c_{mn}(\theta)\frac{\partial a_c^i}{\partial \Psi^k}\right]\frac{\partial \Psi^k}{\partial \theta^b}$$

Since $f^a_{bc}(\theta)$ depend only on θ, the derivative of (3.20) with respect to Ψ^k gives

$$\frac{\partial}{\partial \Psi^k}\left(a_n^i\frac{\partial a_m^j}{\partial \Psi^j}-a_m^j\frac{\partial a_m^i}{\partial \Psi^j}\right)=f^c_{mn}(\theta)\frac{\partial a_c^i}{\partial \Psi^k}$$

so that the right-hand side of the previous expression vanishes and consequently

$$a_c^i\frac{\partial f^c_{mn}(\theta)}{\partial \theta^b}=0$$

Since the matrix a^i_b has rank less than or equal to the smallest value between N and n, it follows that

$$\frac{\partial f^c_{mn}(\theta)}{\partial \theta^a} = 0$$

showing that $f^b_{mn}(\theta)$ are in fact constants. Replacing this result in (3.13) we obtain the Lie theorem

$$[X_a, X_b] = f^c_{ab} X_c \tag{3.21}$$

This implies that the space generated by $\{X_a\}$ defines an algebra with the product (the *Lie product*) defined by the commutator. The result is called *Lie algebra* of the group G denoted by $\mathscr{G}$. The constants f^c_{ab} are antisymmetric in the lower indices $f^c_{ab} = -f^c_{ba}$.

Another property of the Lie algebra is that it is non-associative. Instead of the associativity, it satisfy the *Jacobi identity*

$$[[X_c, X_a], X_b] + [[X_b, X_c], X_a] + [[X_a X_b], X_c] = 0$$

or in terms of the structure constants

$$f^p_{ca} f^n_{pb} + f^p_{bc} f^n_{pa} + f^p_{ab} f^n_{pc} = 0 \tag{3.22}$$

One interesting aspect of Lie's theorem is that almost everything can be done within the Lie algebra, including the representations of the Lie group [62]. This is a consequence of the analytical property which implies that it is possible to recover the full group starting from the Lie algebra.

Theorem 3.2 (The Inverse of Lie Theorem) *A finite transformation of a Lie group G can be obtained from the converging series of infinitesimal transformations generated by its Lie algebra $\mathscr{G}$.*

In fact, consider a set of constants f^c_{ab} satisfying (3.19) and (3.20). This means that there are functions $\mathscr{F}^{-1a}{}_b$ and a^i_a satisfying the equations

$$\begin{cases} \dfrac{\partial \mathscr{F}^{-1b}{}_a}{\partial \theta^c} - \dfrac{\partial \mathscr{F}^{-1b}{}_c}{\partial \theta^a} = f^b_{mn} \mathscr{F}^{-1m}{}_c \mathscr{F}^{-1n}{}_a \\ a^j_m \dfrac{\partial a^i_n}{\partial \Psi^j} - a^j_n \dfrac{\partial a^i_m}{\partial \Psi^j} = f^c_{mn} a^i_c \end{cases} \tag{3.23}$$

Replacing $\mathscr{F}_b^{-1a}$ from (3.15) and applying the initial condition (corresponding to the identity transformation)

$$\mathscr{F}^a_b(0) = \delta^a_b \tag{3.24}$$

we may solve in principle (3.23) for the functions $\mathscr{F}^{-1b}{}_a$. However, it is easier and more intuitive to change the parameterization of the group to a more convenient one, called the *vector parameterization* defined by

$$\theta^a = s^a \tau$$

whose geometrical interpretation is as follows: each set of values $(\theta^1, \ldots, \theta^N)$ defines a straight line with parameter τ in the space of parameters, passing through the origin and with direction s^a. Then, each transformation of the group corresponds to a point in such line. The identity transformation (conventionally described by $\theta^a = 0$) corresponds to origin $\tau = 0$.

In this new parameterization the operation of the group in a space $\mathscr{V}$ can be described by the *line operator* $S(\tau) = S(s^1\tau, \ldots, s^N, \tau)$ such that for each set of constant values of $s^1, \ldots, s^N$, the operator depends only on τ defined in the straight line. Then, the transformation of the field from $\Psi^i(0)$ to $\Psi^i(\tau)$ can be represented by

$$\Psi^i(\tau) = S(\tau)\Psi^i(0) \tag{3.25}$$

where the operator $S(\tau)$ still needs to be defined. For this, consider the variation of the field along the line

$$\frac{d\Psi^i(\tau)}{d\tau} = \frac{\partial \Psi^i}{d\theta^a}\frac{d\theta^a}{d\tau} = \frac{\partial \Psi^i}{\partial \theta^a} s^a = s^a \mathscr{F}^{-1b}{}_a(s,\tau) X_b \Psi^i$$

where in the last equal sign we have used (3.15). Consequently, the variation of Ψ^α can be expressed as $\partial \Psi^i/\partial \tau = \partial S/\partial \tau \Psi^i(0)$. The derivative of (3.25) compared with the above expression gives a differential equation for $S(\tau)$:

$$\frac{dS(\tau)}{d\tau} = s^a \mathscr{F}^{-1b}{}_a(s,\tau) X_b S(\tau)$$

This equation can be integrated with the boundary condition $S(0) = 1$ at $\tau = 0$, compatible with (3.24), obtaining

$$\left.\frac{dS(\tau)}{d\tau}\right]_{\tau=0} = s^a X_a$$

From the analytic dependence on the parameters it follows that $S(\tau)$ is also analytic in τ, so that it can be represented by a converging positive power series

$$S(\tau) = S(0) + \tau \left.\frac{dS}{d\tau}\right]_{\tau=0} + \cdots$$

or, using the above initial conditions,

$$S(\tau) = 1 + \tau s^a X_a + \cdots$$

Therefore, for each straight line defined by the parameters s^a, we may obtain the finite operation of the group. Then, for other finite group operations we only add the rotations of the straight line around the origin. This completes the theorem.

To obtain the finite transformations of a field Ψ we just apply the operator $S(\tau)$ to $\Psi^i(0)$, obtaining

$$\Psi^i(\tau) = \Psi^i(0) + \tau s^a X_a \Psi^i(0) + \cdots$$

From the above theorems it follows also that the existence of Lie subgroups implies the existence of Lie subalgebras, that can be characterized by the structure constants.

As an example, if H is a subgroup of a Lie group G with p parameters, then the commutator of two linear operators of H belongs to H:

$$[X_a X_b] = f^c_{ab} X_c, \quad c = 1, \ldots, p,$$
$$f^c_{ab} = 0, \quad c = p+1, \ldots, N.$$

In particular, using the structure constants we may characterize Lie invariant subalgebras. From this we may define a simple Lie algebra (when it does not have proper invariant subalgebras) which corresponds to a simple group. A semi-simple Lie algebra (which does not have any Abelian invariant subalgebras) also corresponds to a semi-simple Lie group [63, 64] (these properties make an interesting exercise).

Definition 3.4 (Adjoint Representations of Lie Algebras) A representations of a Lie algebra $\mathscr{G}$ is a homomorphism

$$\mathscr{R} : \mathscr{G} \to \mathscr{G}'$$

where $\mathscr{G}'$ is an algebra of linear operators on a space V, such that

$$\mathscr{R}([X_a, X_b]) = [\mathscr{R}(X_a), \mathscr{R}(X_b)]$$

From the definition of Lie algebra it follows immediately that

$$[\mathscr{R}(X_a), \mathscr{R}(X_b)] = f^c_{ab} \mathscr{R}(X_c) \tag{3.26}$$

A consequence of the inverse theorem of Lie is that the representation of a Lie algebra induces the representation of the corresponding Lie group.

Similar to the representations of groups, the representations of Lie algebras are not unique, as they depend on the choice of the representation space, on their action of the algebra on that space, and finally on the choice of the basis of the representation space. Once the action of the algebra and the representation space is chosen, we may finally take a basis of that space and apply the Lie algebra operators $\mathscr{R}(X_a)$ on that basis:

$$\mathscr{R}(X_a)(\eta_i) = \sum \mathscr{R}(X_a)^j{}_i \eta_j \tag{3.27}$$

where $\mathscr{R}(X_a)^j{}_i$ denote the components of the representation matrix.

One particularly interesting representation of a Lie algebra is defined by the space of the Lie algebra itself, using the basis $\{X_a\}$, the same where the structure constants were defined. In this basis the algebra acts in the following way:

$$\tilde{\mathscr{G}}(X_a)X_b \stackrel{\text{def}}{=} [X_a, X_b] = f_{ab}^c X_c$$

where we have used a special notation $\tilde{\mathscr{G}}$ for this representation, which explicitly tells that the representation space is the space of the Lie algebra itself.

Comparing with (3.27), the matrix elements of the adjoint representation associated with the basis (taking $\eta_i = X_a$) are $\tilde{\mathscr{G}}(X_a)X_b = \tilde{\mathscr{G}}(X_a)^c{}_b X_c$. Therefore, it follows from (3.26) that the matrix elements of the adjoint representation are the structure constants of the Lie algebra

$$\tilde{\mathscr{G}}(X_a)^c{}_b = f_{ab}^c \tag{3.28}$$

In the adjoint representation all relevant group quantities are characterized by the structure constants.

Definition 3.5 (Casimir Operators) Given two operators $A = A^a X_a$ and $B = b^b X_b$ defined in the adjoint representation of a Lie algebra $\mathscr{G}$, we may define the product of the two operators consistently with the Lie algebra product as

$$\tilde{\mathscr{G}}(A)\tilde{\mathscr{G}}(B)X_c = [A, [B, X_c]] = A^a b^b [X_a, [X_b, X_c]] = A^a B^b f_{bc}^m f_{am}^n X_n$$

Since $\{X_a\}$ are linearly independent vectors, the above expression defines a matrix with components

$$(\tilde{\mathscr{G}}(A)\tilde{\mathscr{G}}(B))_c^n = A^a B^b f_{bc}^m f_{am}^n$$

whose trace is

$$tr(\tilde{\mathscr{G}}(A)\tilde{\mathscr{G}}(B)) = \sum_c (\tilde{\mathscr{G}}(A)\tilde{\mathscr{G}}(B))_c^c = A^a B^b f_{bn}^m f_{am}^n$$

This is a symmetric bilinear form which defines a scalar product in the Lie algebra space $<,>: \mathscr{G} \times \mathscr{G} \to I\!R$. It can be written as

$$< A, B >= g_{ab} A^a B^b, \text{ where } g_{ab} = f_{am}^n f_{bn}^m \tag{3.29}$$

This product is called the *Killing form* [65]. When the coefficients g_{ab} define an invertible matrix, we obtain a scalar product defining a metric geometry in the space of the Adjoint representation.

The *Casimir operators* of a Lie algebra are defined in the basis $\{X_a\}$ by

$$
\begin{aligned}
C^2 &= g_{ab}X^a X^b = f_{am}^{n} f_{bn}^{m} X^a X^b \\
C^3 &= f_{am}^{p} f_{bn}^{m} f_{cp}^{n} \\
&\vdots \\
C^k &= f_{a_1 m_1}^{m_k} f_{a_2 m_2}^{m_1} \cdots f_{a_k m_k}^{m_{k-1}} X^{a_1} X^{a_2} \ldots X^{a_k}
\end{aligned}
$$

They are *invariant operators* in the sense that they do not depend on the choice of basis in the Lie algebra.

The importance of the Casimir operators resides in the fact that the classification of the unitary irreducible representations of a Lie algebra (or of a Lie group) is given by the eigenvalues of these operators. In particular, Eugene Wigner showed that in the case of the Poincaré group, there are only two Casimir operators: The eigenvalues of the operator C^2 (the mass operator) acting on a Hilbert space gives the mass of the relativistic particles. On the other hand the eigenvalues of the operator C^3 (the spin operator) gives the spins of these particles [30]. This result provided a deep insight into the structure of the physical manifold.

The spectrum of the eigenvalues of the spin operator is discrete (formed by integers and semi-integers). On the other hand the spectrum of the mass operator is continuous, with isolated points (that is, not all real values appear).

Chapter 4
The Algebra of Observables

Observables are measurable quantities such as mass, energy, momentum, spin, and other parameters which are defined in the physical manifold. Together with the notions of physical space and of symmetry groups, the observables and the conditions of observations complete the basic elements of a physical theory.

The observables themselves are usually invariant quantities in the sense that they can be measurable by different observers in different times. In this chapter we study the mathematical structure of observables, which may be in general derived from a tensor structure, but which is usually described as scalar, vector, and spinor fields.

The complete structure of observables depends on the other attributes of a physical theory, including the admissible observers and the definition of the conditions of observations. So, in a sense it requires a complete theory or unified theory. An early version of such theory of everything was proposed in the 1940s by Arthur Eddington under the name of Fundamental Theory. This is an unfinished work, proposing to unify all observable structures in terms of a single algebraic structure that he called the *algebra of observables* [66, 67].

Since all physical fields can be represented as scalars, vectors or spinors, possessing some kind of algebraic structure (defined over some vector space), then in a less pretentious fashion than Eddington's algebra we may describe the observables by means of the *tensor algebra on the physical manifold.*

The scalar fields physical (represented by differentiable functions belonging to an infinite dimensional space) which are usually described as being devoid of algebraic structure can be described as tensors of order zero. We shall see that tensors are a very convenient structure to manifestly display the symmetry properties of these observables.

Tensors appeared as a means to describe the tensions or pressures on the different faces of a crystal or a solid object which requires a different vector field with tangent and normal components to each face. The same is true for any other observable whose existence is detected by some radiation, or by collisions with high energy probes.

Tensors were also known by the name of *absolute differential calculus*. This was essentially due to the fact that the observations must be written in a way that all observers in a theory can recognize the same expression. Thus, the variations of

M.D. Maia, *Geometry of the Fundamental Interactions*,
DOI 10.1007/978-1-4419-8273-5_4,

tensors from one coordinate system to another can be easily detected, regardless of the used reference frame. This implies that not only the object in itself, but the reference frame in which its attributes are defined have a general transformation rule. This is deeply related to the notion of a mathematical analysis that takes into consideration the variations of the observable and the reference frame [68].

We will consider initially tensors defined on the tangent and the cotangent bundles of the physical manifold. In subsequent chapters we will see that these observables may also be defined as objects defined in a more general vector bundle.

4.1 Linear Form Fields

In Chapter 2 we have described tangent vector fields as defined on a tangent bundle on a manifold $\mathcal{M}$. Here we will extend those notions to the corresponding *dual tangent bundle* or *cotangent bundle*.

From linear algebra we have learned that for any given vector space V there is a dual vector space V^*, composed of all linear maps $\varphi : V \to I\!R$. Likewise, a linear form or one-form in an n-dimensional manifold $\mathcal{M}$ is just a map

$$\phi_p : T_p\mathcal{M} \to I\!R$$

such that it is linear on $T_p\mathcal{M}$

$$\phi_p(av_p + bw_p) = a\phi_p(v_p) + b\phi_p(w_p)$$

We may define a sum of linear forms ϕ_p and ψ_p and a product of linear forms by numbers a and b as follows:

$$(a\phi_p + b\psi_p)(v_p) = a\phi(v_p) + b\psi(v_p)$$

Then, it follows from these definitions that the set of all linear forms at a point of $\mathcal{M}$ generates a vector space denoted by $T_p^*\mathcal{M}$, called the *dual tangent space* $T_p\mathcal{M}$.

One basic example of linear form on a manifold is given by the differential of a function f on $\mathcal{M}$, defined by

$$df(v_p) \stackrel{\text{def}}{=} v_p[f] = \sum v^\beta \frac{\partial x^\alpha}{\partial x^\beta} \quad \forall v_p \in T_p\mathcal{M}$$

Taking in particular f to be one of the coordinates of $\mathcal{M}$, x^α, we obtain from the above definition

$$dx^\alpha(v_p) = v_p[x^\alpha] = \sum v^\beta \frac{\partial x^\alpha}{\partial x^\beta} = v^\alpha$$

Therefore, for an arbitrary function f, we may always write

$$df(v_p) = v_p[f] = \sum v^\alpha \frac{\partial f}{\partial x^\alpha}(p) = \sum \frac{\partial f}{\partial x^\alpha}(p) dx^\alpha(v_p)$$

Assuming that this expression holds for any vector v_p at any point p we may just drop the vector and write the result simply as

$$df = \sum \frac{\partial f}{\partial x^\alpha} dx^\alpha$$

which coincides with the usual expression of the differential of a real function on $\mathcal{M}$.

The above example justifies the name *differential form* because a linear form can always be written as the differential of some function: $\phi = df$. It also suggests that the notion of differential of a function can be axiomatized as a linear algebra. It can be argued that such identification is deceiving because we knew beforehand the concept of derivative of f, resulting from the topology of open sets as discussed in Chapter 1.

Another classic example of linear form is given by the scalar product of $\mathcal{M}$ defined by a map $<,>: \ T_p\mathcal{M} \times T_p\mathcal{M} \rightarrow I\!R$ which is bilinear and symmetric. Denoting by $< e_\alpha, e_\beta >= g_{\alpha\beta}$ the components of the scalar product in the basis $\{e_\alpha\}$, it follows that the real function of the tangent space

$$\phi_p = < v_p, \quad >$$

is a linear form because $\phi_p(w_p) = < v_p, w_p >$ is bilinear.

The dual tangent space $T_p^*\mathcal{M}$ is the vector space generated by all linear forms on $T_p\mathcal{M}$. Its dimension is the same as that of $T_p\mathcal{M}$. To see this, consider any linear form (at any point p) ϕ written as the differential of some function f. Applying this form to the vectors of a coordinate basis $\{e_\alpha = \partial/\partial x_\alpha\}$ of $T_p\mathcal{M}$, we obtain

$$\phi(e_\beta) = df(e_\beta) = \sum \frac{\partial f}{\partial x^\alpha} dx^\alpha(e_\beta) = \frac{\partial f}{\partial x^\beta}$$

In particular taking $f = x^\beta$ we obtain

$$dx^\beta(e_\alpha) = \sum \frac{\partial x^\beta}{\partial x^\alpha} = \delta_\alpha^\beta$$

which can be seen as a non-homogeneous linear system of equations in dx^α, for each of the n coordinates of $\mathcal{M}$. To see that these solutions are linearly independent, suppose that we have a null linear combination

$$\sum c_\alpha dx^\alpha = 0$$

Applying to e_β, we obtain the necessary and sufficient condition for linear independency $c_\beta = 0$.

The *dual basis* of $\{e_\alpha\}$ denoted by $\{e^\alpha\}$ (note the position of the index) is defined by the conditions

$$e^\alpha(e_\beta) = \delta^\alpha_\beta$$

In particular $\{dx^\alpha\}$ is the dual basis of the coordinate basis $\{\partial/\partial x^\alpha\}$ of $T_p\mathcal{M}$.

The existence of dual vector spaces in a physical manifold has a profound implication to the structure of the fundamental interactions, although this is not explicit. This duality means that any vector in $T_p\mathcal{M}$ can be associated with an one-form of $T^*_p\mathcal{M}$ and vice versa, $v_p \longleftrightarrow \phi_p$ and $\phi_p \longleftrightarrow v_p$. In other words, $T^*_p\mathcal{M}$ is a vector space isomorphic to $T_p\mathcal{M}$.

However, this isomorphism was established by means of the choice of a particular basis. It does not matter which basis we choose, but it holds only with such a choice. Such isomorphism is sometimes called non-natural. In physics we need to make sure that the use of this isomorphism holds independently of the choice of basis.

As an example of the duality, consider the tangent vector $v_p = \sum v^\alpha e_\alpha$ and the function $\phi : T_p\mathcal{M} \rightarrow I\!R$ given by the scalar product

$$\phi_p(e_\alpha) =< v_p, e_\alpha >= \sum v^\beta g_{\beta\alpha} v^\alpha = \sum v_\alpha v^\alpha$$

where we have denoted $v_\alpha = g_{\beta\alpha} v^\beta$. Therefore, we may associate with the basis vector e_α the one-form

$$e^\beta \stackrel{\text{def}}{=} < e_\beta, \;\; >$$

so that $e^\beta(e_\alpha) =< e_\beta, \, e_\alpha >= g_{\alpha\beta}$. Furthermore, if $v = \sum v^\alpha e_\alpha$, then

$$e^\beta(v) =< e_\beta, \, v >= \sum v^\alpha g_{\alpha\beta} = v_\beta$$

Consequently, given the vector $v_p = \sum v^\alpha e_\alpha \in T_p\mathcal{M}$, the associated one-form defined by the scalar product is

$$\phi_p = \sum g_{\alpha\beta} v^\beta e^\alpha = \sum V_\alpha e^\alpha$$

That is, if v_α are the components of v_p, the components of the one-form ϕ_p are $v_\alpha = \sum g_{\alpha\beta} v^\beta$. In the coordinate basis $\{dx^\alpha\}$ the one-form is written as

$$\phi_p = \sum v_\alpha dx^\alpha$$

whereas v^β are the components of v_p in the basis $\{\partial/\partial x^\beta\}$ of $T_p\mathcal{M}$.

Summarizing, the isomorphism between vector and forms is given by the correspondence

$$v_p = \sum v^\alpha \frac{\partial}{\partial x^\alpha} \longleftrightarrow \phi_p = \sum v_\alpha dx^\alpha$$

Definition 4.1 (The Cotangent Bundle) Clearly, we may repeat all that was done with the tangent bundle to a cotangent bundle to a manifold $\mathscr{M}$ defined as the triad

$$(\mathscr{M},\ \pi,\ T^*\mathscr{M})$$

where $T^*\mathscr{M}$ is the collection of all cotangent spaces in all points of $\mathscr{M}$.

Similar to the definition of tangent vector fields we may define a *cotangent vector field* (or a one-form field, or yet a covariant vector field) as a map

$$\phi : \mathscr{M} \rightarrow T^*\mathscr{M}$$

which associates with each point $p \in M$ a linear form $\phi(p) \in \mathscr{M}$.

Recall from Chapter 2 that if we have a map $F : \mathscr{M} \rightarrow \mathscr{N}$ such that $F(p) \rightarrow q$, then we have a linear map, the *derivative map* $F_* : T_p\mathscr{M} \rightarrow T_q\mathscr{N}$. Similarly we may define the *dual derivative map* as the map

$$F^* : T^*_{F(p)}\mathscr{N} \leftarrow T^*_p\mathscr{M}$$

such that for a given vector $v_p \in \mathscr{M}$, we have

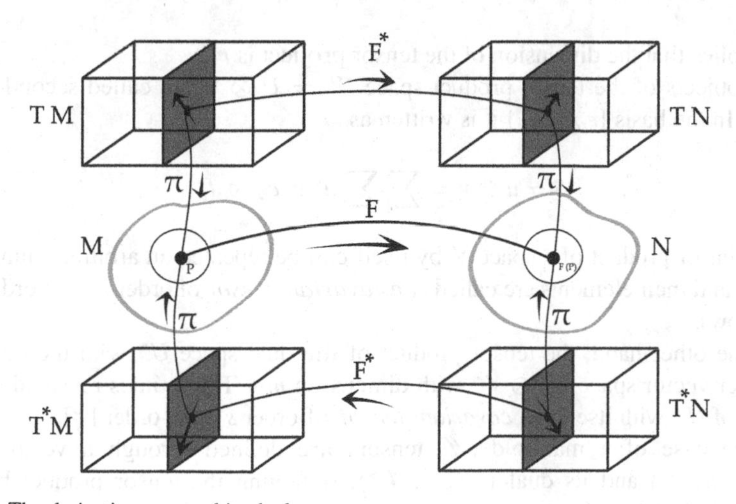

Fig. 4.1 The derivative map and its dual

$$F^*(\phi_{F(p)})(v_p) = \phi(F_* v_p)$$

Notice that the derivative map has the same direction of F, but F^* has the opposite direction as shown in Fig. 4.1. For this reason F^* is frequently referred to as a *pullback map.*

4.2 Tensors

Consider two vector spaces U and V with dimensions m and n, respectively. Then we may define the *Cartesian product* $U \times V$, which is a vector space with dimension $m + n$, composed of all ordered pairs (u, v), $u \in \mathscr{U}$, $v \in \mathscr{V}$. The *tensor product* between U and V is a vector space with dimensions $m \times n$, denoted by $U \otimes V$, such that it associates with the pair $(u, v) \in U \times V$, a vector, denoted by $u \otimes v$, satisfying the following conditions:

(a) The map is bilinear

$$\begin{cases} u \otimes (av + bw) = au \otimes v + bu \otimes w \\ (au + bv) \otimes w = au \otimes w + bv \otimes w \end{cases}$$

(b) If $\{e_\alpha\}$ is a basis of U and $\{f_a\}$ is a basis of W, then $\{e_\alpha \otimes f_a\}$ is a basis of $U \otimes W$.

The bilinear characteristic is consistent with the property that $W = U \otimes V$ is a vector space, so that there are numbers p, q, r, and s, and such that

$$p\, u \otimes u' + qu \otimes v' + rv \otimes u' + sv \otimes v' = (au + bv) \otimes (a'u' + b'v')$$

This implies that the dimension of the tensor product is mn.

The objects of the tensor product space $W = U \otimes V$ are called second-order tensors. In the basis $\{e_\alpha \otimes f_a\}$ it is written as

$$S = u \otimes v = \sum \sum u^\alpha v^a e_\alpha \otimes f_a$$

The tensor product of a space V by itself can be repeated an arbitrary number r of times and their elements are called *contravariant tensor* of order r or of order $[\,^r_0]$ (see below).

On the other hand, the tensor product of the dual space U^* with the dual V^* is another vector space $U^* \otimes V^*$ with dimension mn. The s-times repeated tensor product of U^* with itself is a *covariant tensor* of order s or of order $[\,^0_s]$.

In the case of a manifold $\mathscr{M}$, tensors are defined through a vector bundle $(\mathscr{M}, \pi, \mathscr{V})$ and its dual $(\mathscr{M}, \pi, \mathscr{V}^*)$, obtaining the tensor product bundle $(\mathscr{M}, \pi, \mathscr{V} \otimes \mathscr{V}^*)$.

In particular we may consider the r-repeated tensor product of the *tangent space* $T_p\mathcal{M}$ and the s-repeated tensor product of its dual $T_p^*\mathcal{M}$, obtaining the tangent tensor product bundle $(\mathcal{M}, \pi, \overset{r}{\otimes} T\mathcal{M} \otimes \overset{s}{\otimes} T^*\mathcal{M})$ whose fibers are the spaces

$$T_p\mathcal{M}_s^r = \overset{r}{\otimes} T_p\mathcal{M} \otimes \overset{s}{\otimes} T_p^*\mathcal{M} \tag{4.1}$$

and whose elements are called *mixed tangent tensors* of order $[{}^r_s]$.

By extension of this concept, we define a tensor of order $[{}^0_0]$, or *tensor of order* 0, which is a scalar function or a scalar field.

A tangent tensor of order $[{}^r_s]$ in $\mathcal{M}$ can be written in the coordinate basis $\{e_\mu = \partial/\partial x^\mu\}$ and its dual $\{e^\mu = dx^\mu\}$ as

$$S = S_{\alpha,\ldots,\gamma}{}^{\beta,\ldots,\varepsilon}(e^\alpha \otimes \cdots \otimes e^\gamma) \otimes (e_\beta \otimes \cdots \otimes e_\varepsilon)$$

The collection of all such tensors in all points of $\mathcal{M}$ defines the total space $T_s^r\mathcal{M}$ of a new vector bundle or *tensor bundle* $(\mathcal{M}, \pi, T_s^r\mathcal{M})$. With this bundle we may define a *tensor field* of order $[{}^r_s]$ on $\mathcal{M}$ as a map

$$S : \mathcal{M} \to T_s^r\mathcal{M}$$

such that for each $p \in \mathcal{M}$ it associates a tensor of order $[{}^r_s]$.

The designation of covariant and contravariant tensors is a consequence of the coordinate transformations in tangent spaces. This can be understood from the extension of the map between two tensor bundles.

Theorem 4.1 *Let $F : \mathcal{M} \to \mathcal{N}$ be a diffeomorphism and S an r-contravariant and s-covariant tensor field on $\mathcal{M}$ belonging to $T_p\mathcal{M}_s^r$. Then there is a derivative map F_* for each $T_p\mathcal{M}$ and a "pull back" map F^* for each $T_p\mathcal{M}^*$. These maps induce a dual derivative map*

$$F_*^* : T_p\mathcal{M}_s^r \to T_{F(p)}(\mathcal{N})_s^r$$

given by

$$F_*^*(s) = (s^{\mu,\ldots,\nu}{}_{\rho,\ldots,\sigma})(F_*(e_\mu) \otimes \cdots \otimes F_*(e_\nu)) \otimes (F^*(e^\rho) \otimes \cdots \otimes F^*(e^\sigma)) \tag{4.2}$$

This theorem is a direct extension of the previous definitions of the derivative and the pullback maps to the tensor bundle and needs no further comment. However, in the particular case where $\mathcal{M} = \mathcal{N}$ and where F is a *coordinate transformation* in $\mathcal{M}$, the expressions of $F_*^*(s)$ give the transformation of the components of the tensor field

$$s'^{\alpha,\ldots,\gamma}{}_{\beta,\ldots,\varepsilon} = \frac{\partial x^\mu}{\partial x'^\alpha} \cdots \frac{\partial x^\nu}{\partial x'^\gamma} \cdot \frac{\partial x'^\beta}{\partial x^\rho} \cdots \frac{\partial x'^\varepsilon}{\partial x^\sigma} s^{\rho,\ldots,\sigma}{}_{\mu,\ldots,\nu}$$

As we see, each component of the tensor with upper indices $s^{\rho,\dots,\sigma}{}_{\mu,\dots,\nu}$ transforms with the inverse of the Jacobian matrix $J^{-1} = (\partial x^{\mu}/\partial x'^{\alpha})$ of the coordinate transformation, so that these components are said to be *contravariant*. On the other hand, the lower indices transform with the Jacobian matrix $J = (\partial x'^{\beta}/\partial x^{\rho})$, so that these components are said to be *covariant*. This is a little confusing because the components transform in the inverse sense of the basis. Figure 4.1 may help to make this clearer.

4.3 Exterior Algebra

As we have seen, a one-form in an n-dimensional manifold $\mathcal{M}$ is just a linear map $\phi : T_p\mathcal{M} \to I\!R$. The set of all such one-forms in $T_p\mathcal{M}_n$ is the dual vector space $T_p^*\mathcal{M}$.

A *two-form* in $\mathcal{M}$ is *bilinear and alternate* map

$$\omega : T_p\mathcal{M} \times T_p\mathcal{M} \to I\!R$$

The alternate condition means that $\omega(v_p, w_p) = -\omega(w_p, v_p)$.

The set of all two-forms on $\mathcal{M}$, denoted by $T_p^{*2}\mathcal{M}$, is a vector space with respect to the sum operation

$$(\omega + \omega')(v_p, w_p) = \omega(v_p, w_p) + \omega'(v_p, w_p)$$

and products by real numbers

$$(\alpha\omega)(v_p, w_p) = \alpha(\omega(v_p, w_p))$$

As an example consider two one-forms ϕ and ψ in $\mathcal{M}$. Then, the determinant defined as an operator in $T\mathcal{M} \times T\mathcal{M}$

$$\begin{vmatrix} \phi(v_p) & \phi(w_p) \\ \psi(v_p) & \psi(w_p) \end{vmatrix} = \phi(v_p)\psi(w_p) - \phi(w_p)\psi(v_p)$$

defines a two-form on $\mathcal{M}$ because it is real, bilinear, and alternate.

In particular for $\phi = dx^{\mu}$ and $\psi = dx^{\nu}$, $v_p = e_{\rho} = \partial/\partial x^{\rho}$ and $w_p = e_{\sigma} = \partial/\partial x^{\sigma}$ we obtain

$$\begin{vmatrix} dx^{\mu}(e_{\rho}) & dx^{\mu}(e_{\sigma}) \\ dx^{\nu}(e_{\rho}) & dx^{\nu}(e_{\sigma}) \end{vmatrix} = \delta^{\mu}_{\rho}\delta^{\nu}_{\sigma} - \delta^{\mu}_{\sigma}\delta^{\nu}_{\rho} \tag{4.3}$$

This result suggests that the symbolic determinants, where its entries are operators on pairs of vectors (v, w) and they should not be regarded as numbers (otherwise they would be zero),

$$\begin{vmatrix} dx^\mu & dx^\mu \\ dx^\nu & dx^\nu \end{vmatrix} \tag{4.4}$$

are linearly independent. To see this, consider the coefficients $c_{\mu\nu}$ such that

$$\sum c_{\mu\nu} \begin{vmatrix} dx^\mu & dx^\mu \\ dx^\nu & dx^\nu \end{vmatrix} = 0$$

Applying to the pair of the coordinate basis vectors (e_ρ, e_σ) and remembering the dual basis definition $dx^\mu(e_\rho) = \delta^\mu_\rho$ it follows from (4.3) that

$$\sum c_{\mu\nu} \left(dx^\mu(e_\rho)dx^\mu(e_\sigma) - dx^\mu(e_\sigma)dx^\nu(e_\rho)\right) = c_{\rho\sigma} - c_{\sigma\rho} = 0$$

Due to the alternate condition, only the anti-symmetric components of $c_{\rho\sigma}$ survive in the above sum. Therefore $c_{\rho\sigma} - c_{\sigma\rho} = 2c_{\rho\sigma} = 0$ which means that determinants (4.4) are linearly independent.

Actually any two-form can be written as a linear combination of the two-forms (4.4). Let ω be an arbitrary two-form. Then $\omega(e_\mu, e_\nu) = 2a_{\mu\nu}$ are real numbers and with them we construct another two-form:

$$\psi = \sum a_{\mu\nu} \begin{vmatrix} dx^\mu & dx^\mu \\ dx^\nu & dx^\nu \end{vmatrix}$$

It follows that

$$\psi(e_\rho, e_\sigma) = \sum a_{\mu\nu} \begin{vmatrix} dx^\mu & dx^\mu \\ dx^\nu & dx^\nu \end{vmatrix} (e_\mu, e_\nu) = a_{\mu\nu} \begin{vmatrix} dx^\mu(e_\rho) & dx^\mu(e_\sigma) \\ dx^\nu(e_\rho) & dx^\nu(e_\sigma) \end{vmatrix} = \omega(e_\rho, e_\sigma)$$

so that $\psi = \omega$. The properties of the symbolic determinant (4.4) suggest the notation

$$dx^\mu \wedge dx^\nu \stackrel{\text{def}}{=} \begin{vmatrix} dx^\mu & dx^\mu \\ dx^\nu & dx^\nu \end{vmatrix} \tag{4.5}$$

so that any two-form can be written as

$$\omega = \sum \omega(e_\mu, e_\nu) dx^\mu \wedge dx^\nu$$

Consequently $\{dx^\mu \wedge dx^\nu_{\mu\nu}\}$ defines a basis of the space of two-forms $T_p^{2*}\mathcal{M}$. Taking the dimension of the manifold $\mathcal{M}$ to be n, the alternate condition gives a total of $n(n-1)/2$ independent elements in that basis, so that it has $n(n-1)/2$ dimensions.

Definition 4.2 (k-Forms) The concept of two-forms extends naturally to *k-forms* in $\mathcal{M}$ as a map

$$\xi : T_p\mathcal{M} \times T_p\mathcal{M} \times \cdots \times T_p\mathcal{M} \to I\!R \quad (k\text{-factors})$$

such that it is *k-linear* and *alternate*.

The alternate condition means that for any set of indices $\{\mu_1, \mu_2, \ldots, \mu_k\}$, we have

$$\xi(e_{\mu_1}, \ldots, e_{\mu_k}) = \varepsilon^{1,\ldots,k}_{\mu_1,\ldots,\mu_k} \xi(e_1, \ldots, e_k)$$

where

$$\varepsilon^{1,\ldots,k}_{\mu_1,\ldots,\mu_k} = \begin{cases} +1 \text{ if } \mu_1, \ldots, \mu_k \text{ is an even permutation of } 1, \ldots, k \\ -1 \text{ if } \mu_1, \ldots, \mu_k \text{ is an odd permutation of } 1, \ldots, k \\ \;\;0 \;\text{ in all other cases} \end{cases} \tag{4.6}$$

is a generalization of the Levi-Civita permutation symbol to k dimensions. The set of all such k-forms endowed with the sum and multiplication operations by numbers defines a vector space $T_p^{k*}\mathcal{M}$.

Again, we may use the determinant polynomial structure to generate k-forms. For example, if ϕ, ψ, and ζ are three-forms, then the determinant

$$\begin{vmatrix} \phi(u_p) & \phi(v_p) & \phi(w_p) \\ \psi(u_p) & \psi(v_p) & \psi(w_p) \\ \zeta(u_p) & \zeta(v_p) & \zeta(w_p) \end{vmatrix} = \phi \wedge \psi \wedge \zeta(u_p, v_p, w_p)$$

is trilinear and alternate.

Likewise, using the same arguments for two-forms, the $k \times k$ symbolic determinants are k-forms

$$dx^{\mu_1} \wedge \cdots \wedge dx^{\mu_k} = \begin{vmatrix} dx^{\mu_1} & dx^{\mu_1} & \cdots & dx^{\mu_1} \\ dx^{\mu_2} & dx^{\mu_2} & \cdots & dx^{\mu_2} \\ \vdots \\ dx^{\mu_k} & dx^{\mu_k} & \cdots & dx^{\mu_k} \end{vmatrix}$$

They define a base of the space of k-forms denoted by $T_p^{k*}\mathcal{M}$. Thus, any k-form can be written as

$$\xi = \sum \xi_{\mu_1 \mu_2 \cdots \mu_k} dx^{\mu_1} \wedge \cdots \wedge dx^{\mu_k}$$

Notice that the number of k-forms which are linearly independent depends on the dimensions n of the manifold $\mathcal{M}$. For example,

If $n = 3$ and $k = 1$ we have three independent two-forms.
If $n = 3$ and $k = 2$ we have again three independent two-forms.
If $n = 3$ and $k = 3$ we have just one three-form.

To see the last case in more detail, consider the basis of $T_p^{3*}\mathcal{M}$ given by the determinant

$$dx^{\mu} \wedge dx^{\nu} \wedge dx^{\rho}(e_1, e_2, e_3) = \begin{vmatrix} \delta_1^{\mu} & \delta_2^{\mu} & \delta_3^{\mu} \\ \delta_1^{\nu} & \delta_2^{\nu} & \delta_3^{\nu} \\ \delta_1^{\rho} & \delta_2^{\rho} & \delta_3^{\rho} \end{vmatrix}, \qquad \mu, \nu, \rho = 1, \ldots, 3$$

It is clear that this is different from zero only when $\mu \neq \nu \neq \rho$, so that we have in fact only one element in the basis.

On the other hand, if $n = 3$ and $k > 3$ the basis elements will have always two equal indices, so that they vanish. Therefore all k-forms in an n-dimensional manifold, with $k > n$, are necessarily zero.

In general we can summarize this result as

$$\dim T_p^{*k}\mathcal{M} = C_n^k = \frac{n(n-1)(n-2)\cdots(n-k+1)}{k}$$

Definition 4.3 (The Exterior Product) The properties of determinant (4.5) convey naturally to a product operation as follows: Given the spaces $T_p^{*k}\mathcal{M}$ and $T_p^{*\ell}\mathcal{M}$, the *exterior product* between them is a map

$$\wedge : T_p^{*k}\mathcal{M} \times T_p^{*\ell}\mathcal{M} \to T_p^{*k+\ell}\mathcal{M}$$

which associates with the pair (ϕ, ψ)

$$\phi = \sum a_{\mu_1,\ldots,\mu_k} dx^{\mu_1} \wedge \cdots \wedge dx^{\mu_k}$$
$$\psi = \sum b_{j_1,\ldots,j_\ell} dx^{j_1} \wedge \cdots \wedge dx^{j_\ell}$$

a $(k+\ell)$-form $\phi \wedge \psi$ given by

$$\phi \wedge \psi = \sum c_{\mu_1,\ldots,\mu_{k+\ell}}\, dx^{\mu_1} \wedge \cdots \wedge dx^{\mu_{k+\ell}}$$

As an example consider a three-dimensional manifold $\mathcal{M}$. The exterior product of the one-form

$$\phi = 3dx^1 + x^2 dx^3 \in T_p^* \mathcal{M}$$

with the two-form

$$\psi = 4xzdx^1 \wedge dx^2 \in T_p^{*2} \mathcal{M}$$

is the three-form on $\mathcal{M}$ given by

$$\phi \wedge \psi = 4zx^3 dx^1 \wedge dx^2 \wedge dx^3 \in T_p^{*3} \mathcal{M}$$

From the properties of forms in three-dimensional spaces it follows that this three-form is isomorphic to a one-form.

Exercise 4.1 As an exercise show that if ϕ is a k-form, ψ an ℓ-form, ζ an m-form, and f a scalar function defined on an n-dimensional manifold $\mathcal{M}$, then the exterior product satisfies the properties

(a) Associativity: $(\phi \wedge \psi) \wedge \zeta = \phi \wedge (\psi \wedge \zeta)$
(b) Anticommutativity: $\phi \wedge \psi = (-1)^{k+l} \psi \wedge \phi$
(c) Distributivity: $\phi \wedge (\psi + \zeta) = \phi \wedge \psi + \phi \wedge \zeta$ (for $\ell = m$)
(d) Product by scalar functions: $f \wedge \psi = f\psi$

The inclusion of the property (d) and the definition of tensors of order $[^0_0]$ make the exterior algebra on a manifold $\mathcal{M}$ consistent with the tensor algebra obtained with all skew-symmetric tensors in $\mathcal{M}$.

For example the exterior product of two one-forms written as

$$\phi \wedge \psi = \sum w_{\mu\nu} dx^\mu \wedge dx^\nu$$

corresponds to a rank-2 anti-symmetric tensor:

$$w = w_{\mu\nu}(e^\mu \otimes e^\nu - e^\nu \otimes e^\mu)$$

and vice versa: *Given an anti-symmetric rank-2 tensor, it corresponds to a two-form, which corresponds to the exterior product of two one-forms.*

Exercise 4.2 Show that the vector product of $I\!R^3$ corresponds to the exterior product of the two corresponding one-forms.

From the above properties we may infer that in principle we could use just the tensor algebra instead of detailing the exterior algebra as we did. However, the exterior product allows us to write the operations of anti-symmetric tensors in a more compact form and easier to interpret. For that reason it is convenient to use the exterior algebra concomitantly with the tensor algebra.

For example, the surface integral of an anti-symmetric tensor (where the index anti-symmetrization is indicated by the square brackets within the indices)

$$\int\int T_{[\mu\nu]}dx^{\mu}dx^{\nu}$$

is equivalent to the surface integral of a two-form $\int\int T_{\mu\nu}dx^{\mu}\wedge dx^{\nu}$.

Definition 4.4 (k-Form Fields) Given the cotangent bundle with total space $T^{*k}\mathcal{M}$, a *k-form field* in $\mathcal{M}$ is a map

$$\omega : \mathcal{M} \rightarrow T^{k*}\mathcal{M}$$

such that it associates with each point of $\mathcal{M}$ a k-form $\omega(p)$. In coordinate basis this k-form field can be expressed as

$$\omega = \sum \omega_{\mu_1,\ldots,\mu_k}dx^{\mu_1}\wedge\cdots\wedge dx^{\mu_k}$$

When the components $\omega_{\mu_1,\ldots,\mu_p}$ are differentiable functions on $\mathcal{M}$ then ω is called a differentiable k-form field.

Definition 4.5 (Exterior Derivative) Let ω be a k-form of $T_p^{*k}(\mathcal{M})$ written in coordinate basis as

$$\omega = \sum \omega_{\mu_1,\ldots,\mu_k}dx^{\mu_1}\wedge\cdots\wedge dx^{\mu_k}$$

The *exterior derivative* $d\omega$ of ω is a map

$$d\ :\ T_p^{*k}(\mathcal{M}) \rightarrow T_p^{*k+1}(\mathcal{M})$$

which associates with ω a $(k+1)$-form given by

$$d\omega = \sum(d\omega\mu_{\mu_1,\ldots,\mu_k})\wedge dx^{\mu_1}\wedge\cdots\wedge dx^{\mu_k}$$

where $d\omega_{\mu_1,\ldots,\mu_k}$ is the differential of each scalar component as a scalar field.

The exterior derivative satisfies the following properties (exercise):

(a) The exterior derivative of a zero-form $f(x)$ is the one-form given by the differential $df(x)$.
(b) If ϕ and ψ are k-forms, then $d(a\phi + b\psi) = a\,d\phi + b\,d\psi \quad a, b \in I\!R$.
(c) If f is a zero-form, then $d(f\phi) = df \wedge \phi + f d\phi$.
(d) If ϕ is a k-form and ψ is an ℓ-form, then

$$d(\phi \wedge \psi) = d\phi \wedge \psi - (-1)^{(k+\ell)} \phi \wedge d\psi$$

As an example, consider the exterior derivative of the one-form $\phi = \sum \phi_\mu dx^\mu$, obtaining the two-form

$$d\phi = \sum d\phi_\mu \wedge dx^\mu = \sum \frac{\partial \phi_\mu}{\partial x^\nu} dx^\nu \wedge dx^\mu$$

As we have seen, all $(n+1)$-forms defined on an n-dimensional manifold vanish, so that the exterior derivative of an n-form in an n-dimensional manifold necessarily vanishes.

Since the coordinates x^μ are zero-forms in $\mathcal{M}$, their exterior derivatives are one-forms dx^μ defining a basis of $T_p^* M$, which is the dual basis of $T_p M$.

Example 4.1 (Curl of a Vector) Consider a one-form $\phi = \sum \phi_\mu \, dx^\mu$ in $I\!R^3$. The exterior derivative of ϕ is a two-form

$$d\phi = \sum d\phi_\nu \wedge dx^\nu = \sum \frac{\partial \phi_\nu}{\partial x^\mu} dx^\mu \wedge dx^\nu$$

which has three independent components, so that it is equivalent to some one-form, which is associated with a vector $\mathbf{v} = \sum \phi^\mu (\partial / \partial x^\mu)$, with components given by

$$d\phi = \left(\left(\frac{\partial \phi_1}{\partial x^2} - \frac{\partial \phi_2}{\partial x^1} \right), \left(\frac{\partial \phi_1}{\partial x^3} - \frac{\partial \phi_3}{\partial x^1} \right), \left(\frac{\partial \phi_2}{\partial x^3} - \frac{\partial \phi_3}{\partial x^2} \right) \right)$$

In other words, the exterior derivative of ϕ is equivalent to the rotational $\nabla \times V$.

Chapter 5
Geometry of Space–Times

The concept of space–time is discussed in several publications on philosophy and on the foundations of physics [69, 70]; here we are interested in knowing how the physics of space–time determines the geometry. In this, we were much influenced by Penrose's writings and lectures [71].

As we have detailed in the introduction, we take the physical manifold as the space of perceptions in the sense of Kant. These perceptions are in one way or another associated with a physical interaction, from which we eventually extract the geometry. Thus, *a space–time is the physical manifold endowed with a notion of geometry determined by a physical process.* As it is evident, the observational methods evolve with technology, so that the concept of space–time is not static. In the following we review this process of evolution from Galilei to Einstein's general relativity. Later on we discuss some future perspectives.

A first point to be made clear is that *as a set of observers and observables all space–times are four dimensional.* This is consistent with the structure of Maxwell's equations describing the electromagnetic field. This will be made clear later on, and it is a consequence of the dual properties of electromagnetic field, and in general of the Yang–Mills field, from the point of view of both experimental evidences and mathematical consistency. On the other hand, we shall see also that the gravitational interactions in the sense of Einstein do not have the same gauge structure and therefore they do not have the same dimensionality limitations.

5.1 Galilean Space–Time

The *"Galilean" space–time* which we will denote by $\mathscr{G}_4$ is a four-dimensional manifold, with a geometry and symmetry defined by the *motion of a free particle, along geodesics* defined by Newton's first law and the absolute time.

The notion of absolute time introduced by Newton in his *Principia* reads as [72] follows: *"Absolute, universal, true and mathematical time exists by itself and flows equably, without relating to anything external."*

M.D. Maia, *Geometry of the Fundamental Interactions*,
DOI 10.1007/978-1-4419-8273-5_5, © Springer Science+Business Media, LLC 2011

Unlike the three space-like coordinates, we have no control of the absolute time (*it exists by itself*). The absoluteness of time means that it is independent of the position of the observer (*it flows equably*). Therefore, it follows from the above definition that absolute time transforms only in scale and origin. That is, if t is an absolute time, then

$$t' = at + b, \quad a \neq 0$$

where a and b are constants, is another absolute time. From this transformation property it follows that the absolute time can be regarded as a regular function (its derivative never vanishes) and so it serves as a kind of special coordinate, with an arbitrary non-zero scale (a) and an arbitrary origin (b). Since the absolute time transforms only into another absolute time, the Galilean space–time has a "*product topology*," characterized by

$$\mathcal{G}_4 = I\!R^3 \times I\!R$$

where $I\!R$ represents the absolute time axis. The topological product means that time never mixes with the three coordinates in $I\!R^3$. Therefore, *the Galilean space–time is a four-dimensional physical manifold with absolute time*.

As we all know, the absolute time implies the existence of some kind of instantaneous communication between all observers belonging to the same space-like hypersurface defined for a value $t = t_0$, regardless of where they are. The concept of "*distant instantaneous interaction*" presented a major difficulty for the development of physics, something that was only solved with the special theory of relativity in the beginning of the 20th century.

A *simultaneity section* at t_0 is a submanifold of $\mathcal{G}_4$ defined at the instant $t = t_0$ by the set

$$\Sigma_{t_0} = \{p \in \mathcal{G}_4 \mid t(p) = t_0\}$$

One interesting consequence of this definition is that two simultaneous sections do not intersect or else they coincide.

Indeed, suppose that we have a point p_0 common to two simultaneous sections: $p_0 \in \Sigma_{t_1}$ and $p_0 \in \Sigma_{t_2}$. Then $t(p_0) = t_1$ and $t(p_0) = t_2$ so that $t_1 = t_2$ and from the definition it follows that

$$\Sigma_{t_1} = \{p \in \mathcal{G}_4 \mid t(p) = t_1\} = \{p \in \mathcal{G}_4 \mid t(p) = t_2\} = \Sigma_{t_2}$$

We say that the absolute time induces a *foliation* of the Galilean space–time, where each "leaf" is a simultaneity section Σ_t. In other words, *the Galilean space–time is the total space of a vector bundle*, where the fibers are simultaneity sections and the base manifold is the time axis as shown in Fig. 5.1.

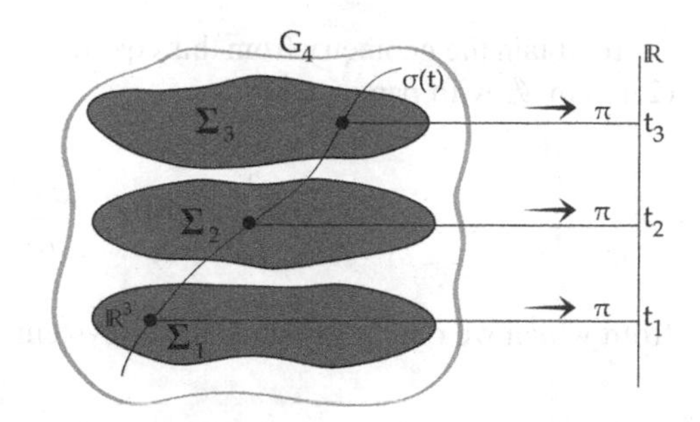

Fig. 5.1 The Galilean space–time $\mathscr{G}_4$

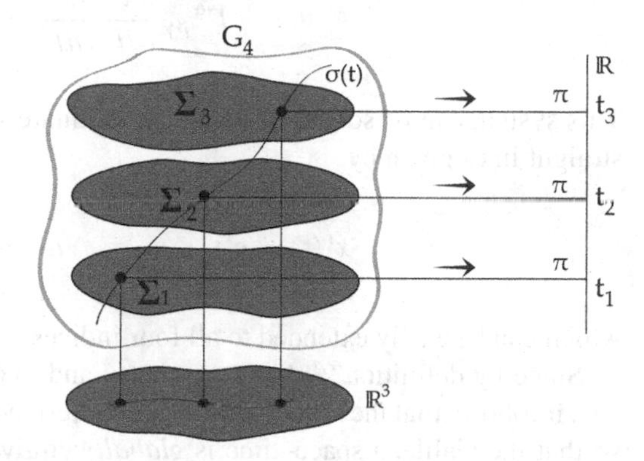

Fig. 5.2 Projecting the Galilean space–time on $I\!R^3$

The four-dimensional curve γ shown in Fig. 5.2 represents the time and space evolution or the *world-line* of an *event* in space–time $\mathscr{G}_4$. The product topology allows us to write the Galilean space–time as if the particles in motion are points of $I\!R^3$ written as coordinates depending on the absolute time. This is pictured in the projection $I\!R^3$ of Fig. 5.2 where the world-lines are projected as three-dimensional *trajectories* in a special three-dimensional manifold.

As already commented the observables have mass, charge, spins, and energy, so that they cannot be identified with the points in the parameter space $I\!R^3(t)$ of the simultaneous sections. In spite of this difference, by abuse of language we refer to these as point particles in $I\!R^3$. It is in this projection that Newton's first law was established as

$$\frac{d^2x^i(t)}{dt^2} = 0, \quad i = 1, \ldots, 3 \tag{5.1}$$

Noting that $d^2t/dt^2 = 0$, (5.1) can be extended to the four-dimensional Galilean physical space–time as

$$\frac{d^2x^\alpha}{dt^2} = 0, \quad \alpha = 1, \ldots, 4$$

where we have identified $x^4 = t$.

To obtain the geometry from this equation, compare it with the geodesic equation (2.15) in $\mathcal{G}_4$ with parameter t,

$$\frac{d^2x^\alpha}{dt^2} + \Gamma^\alpha{}_{\beta\gamma}\frac{dx^\beta}{dt}\frac{dx^\gamma}{dt} = 0$$

from which we obtain a homogeneous system of equations on $\Gamma^\alpha{}_{\beta\gamma}$

$$\Gamma^\alpha{}_{\beta\gamma}\frac{dx^\beta}{dt}\frac{dx^\gamma}{dt} = 0 \tag{5.2}$$

This system can be solved by using a coordinate system in which the geodesics are straight lines given by

$$x^i(t) = a^i t + p^i, \qquad a^i, p^i \text{ constants}$$

which can be easily extended to all four indices.

Since by definition $\mathcal{G}_4$ has a geometry and symmetry defined by Newton's first law, it follows that the space–time is parameterized by such coordinates everywhere, so that the Galilean space–time is *globally* equivalent to $\mathbb{R}^3 \times \mathbb{R}$. The illusion is almost complete and we all think of the first Newton's law as established for point particles in $\mathbb{R}^3 \times \mathbb{R}$.

Finally note that (5.1) does not have the same form in different coordinate systems. For example, for an observer sitting in a carrousel in motion there will be an additional centrifugal force attached to (5.1), with respect to an observer sitting on the ground. Therefore, (5.1) will look different for the two observers. The character of "law" attached to (5.1) means that the equation must be understood equally by the observers who agree with that same expression. The observer sitting on the carrousel will not agree with the one sitting on the ground. This means that (5.1) is not invariant under an arbitrary coordinate transformation of $\mathcal{G}_4$, but only to the coordinate transformations belonging to the symmetry of (5.1).

Suppose that we have a transformation $x^i \rightarrow x'^i$, such that $d^2x^i/dt^2 = 0$ and $d^2x'^i/dt^2 = 0$. We find that the equation is not the same for an arbitrary transformation, but only those like (exercise)

$$\begin{cases} x^i = \sum a^i_j x_j + c^i t + d^i, \quad A = \text{matrix}(a^i_j), \quad AA^T = 1, \quad c^i, d^i = \text{constants} \\ t' = at + b, \quad a, b = \text{constants} \quad a \neq 0 \end{cases} \tag{5.3}$$

where A^T is the transpose of A, the set of such transformations define a group with respect to the composition of transformations. It is known as the *restricted Galilean group* (or simply the Galilean group).

5.2 Newton's Space–Time

The Newtonian space–time denoted here by $\mathcal{N}_4$ is a four-dimensional manifold with absolute time, in which the trajectory of a particle with mass m, under the influence of gravitation only (a free falling particle), is a geodesic defined by Newton's gravitational law:

$$F = G\frac{mm'}{r^2} \tag{5.4}$$

To compare with the geodesic differential equation, let us first write (5.4) as a differential equation.

Consider a unit test particle under the influence of the gravitational field of a spherically distributed infinitesimal mass $m' = dm$ with uniform density ρ. Using spherical coordinates (r, θ, φ) we may write the elementary mass of each shell as

$$dm = \rho\, r^2 \sin\theta\, dr\, d\theta\, d\varphi$$

Assuming that Newton's gravitational force derives from a time-independent potential ϕ, $F = -\nabla\phi$, we obtain by integrating on the spherical solid angle 4π, it follows that

$$\nabla^2\phi = -4\pi G\rho \tag{5.5}$$

which is Poisson's equation for the Newtonian gravitational potential. In the left-hand side we have applied the divergence theorem with a minus sign to account for the conventional outward orientation of the sphere.

The geodesic equation of motion of the free falling test particle with unit mass is obtained by replacing the gravitational force for a unit mass by $F = -\nabla\phi = a(t)$. This is a three-dimensional expression but since $d^2t/dt^2 \equiv 0$ and using the fact that ϕ is not time dependent, we may use again the notation $x^4 = t$ and write the geodesic equation of motion as a four-dimensional equation

$$\frac{d^2x^\alpha}{dt^2} = -\frac{\partial\phi}{\partial x^\alpha}$$

Comparing this equation with the geodesic equation written with the same absolute time parameter t

$$\frac{d^2x^\alpha}{dt^2} + \Gamma^\alpha{}_{\beta\gamma}\frac{dx^\beta}{dt}\frac{dx^\gamma}{dt} = 0$$

we obtain

$$\Gamma^\alpha{}_{\beta\gamma}\frac{dx^\beta}{dt}\frac{dx^\gamma}{dt} = \frac{\partial\phi}{\partial x^\alpha} \tag{5.6}$$

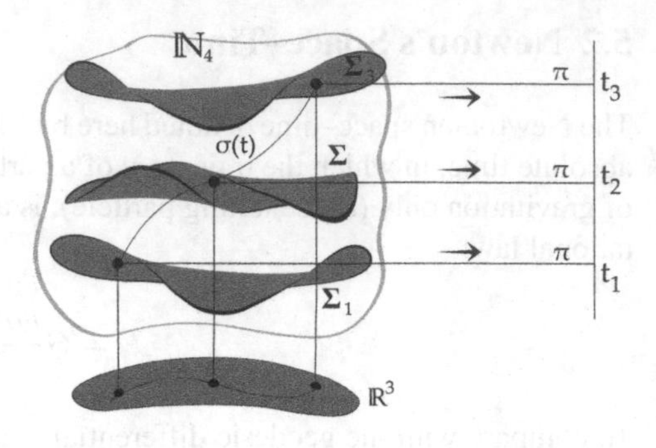

Fig. 5.3 The Newtonian space–time

so that the affine connection coefficients $\Gamma^{\alpha}{}_{\beta\gamma}$ are determined by the gravitational potential ϕ.

Recalling that the absolute time implies the existence of simultaneity sections, like in the Galilean case, the simultaneity sections do not intersect, but they are not necessarily flat because the geodesic lines are not necessarily straight lines as illustrated in Fig. 5.3. Since each of these simultaneity sections is a hypersurface of $\mathcal{N}_4$ defined by a value of t, they may be defined by an equation like

$$t(x^1, x^2, x^3) = \text{constant}$$

These are time-oriented surfaces, with normal vectors $\eta_\alpha = \partial t/\partial x^\alpha$. Therefore, (5.6) may be solved in terms of the normal vectors to give

$$\Gamma^{\alpha}{}_{\beta\gamma}(\phi) = \frac{\partial \phi}{\partial x^\gamma}\eta_\alpha \eta_\beta \tag{5.7}$$

showing how the Newtonian gravitational potential determines the connection. The three-dimensional projection is also shown.

Similar to the case of the Galilean space–time, this determines an affine geometry in $\mathcal{N}_4$.

5.2.1 The Curvature of Newton's Space–Time

As we see in Chapter 2, the existence of a connection determines a curvature tensor. Therefore, the connection determined by the Newtonian gravitational potential must also have a Riemann-like curvature.

For that purpose consider two free falling particles a and b along two neighboring geodesics in $\mathcal{N}_4$, under the influence of Newton's gravitational potential satisfying Poisson's equation (5.5).

Along this fall, the particles exchange signals to one another, with speed P, traveling through the smallest distance between them, so that in each simultaneity section they follow a geodesic (the signal geodesic), such that when the particle a

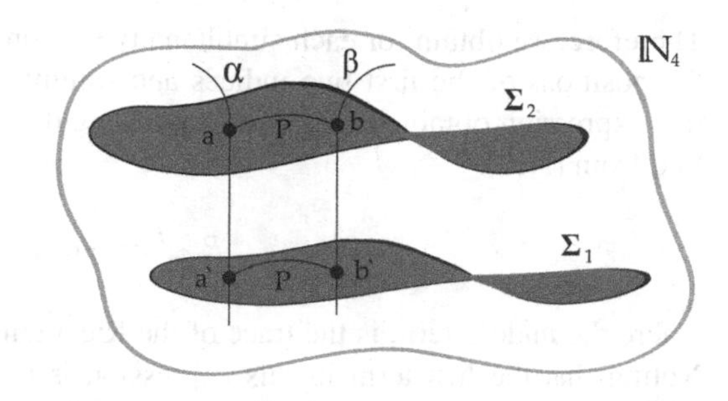

Fig. 5.4 Geodesic deviation in $\mathcal{N}_4$

reaches the position a′, the particle b will reach the position b′ (Fig. 5.4), in such a way that we obtain a closed, four-sided geodesic figure, satisfying the equations

$$\nabla_T T = 0 \quad \text{the fall geodesics} \tag{5.8}$$

$$\nabla_P P = 0 \quad \text{the signal geodesics} \tag{5.9}$$

$$\nabla_T P = \nabla_P T \quad \text{the closure condition} \tag{5.10}$$

Replacing these conditions in the expression of the Riemann tensor (2.5) we obtain

$$R(T, P)T = \nabla_T \nabla_P T = \nabla_T \nabla_T P$$

Since P is velocity vector of the signal, its time variation must equal the fall acceleration of fall a:

$$\nabla_T P = a = \text{Signal acceleration during the fall} \tag{5.11}$$

On the other hand, we obtain the same Riemann tensor by reversing the path order so that $R(T, P)P = -R(T, P)T$. Therefore, from the closure condition we obtain

$$R(T, P)T = -\nabla_P \nabla_T P = -\nabla_P\, a \tag{5.12}$$

which describes the geodesic deviation with respect to the Newtonian connection defined by the Newtonian potential (5.7).

To see the relation of the above curvature to the Newtonian gravitational field, consider a coordinate basis, where the tangent vector T to the fall geodesic coincides with the time base vector which we denote by e_4. In this case, $T^\alpha = \delta^\alpha_4$. On the other hand, take the tangent vector P to the signal geodesic tangent to a simultaneity section, aligned to any spatial base vector e_i, $i = 1, \ldots, 3$. Remembering from Chapter 2 that in a general basis the components of the Riemann tensor are given by $R(e_\mu, e_\nu)e_\rho = R_{\mu\nu\rho}{}^\sigma e_\sigma$, we obtain in our coordinate basis

$$R(e_4, e_i)e_4 = R_{4i4}{}^\sigma e_\sigma = -(\nabla_{e_i} a)^\sigma e_\sigma = a^\sigma{}_{,i} e_\sigma \tag{5.13}$$

Therefore, we obtain for each simultaneity section $R_{4i4}{}^{\alpha} = a^{\alpha}{}_{,i}$. After exchanging the positions of the first two indices and summing the three-dimensional trace of this expression obtained for $\sigma = i$, and using the definition of the Ricci tensor, we find from (5.12)

$$R_{i44}{}^{i} = -a^{i}{}_{,i}$$

where the middle term is the trace of the Ricci tensor of each simultaneity section. Noting that the last term in this expression is the divergence of the acceleration of the falling particle, we obtain for a particle of unit mass the three-dimensional divergence of the Newtonian equation

$$a^{i}{}_{,i} = < \nabla, a > = < \nabla, F > = -\nabla^2\phi = 4\pi G\rho$$

which is equivalent to the geometric equation

$$R_{44} = -4\pi G\rho \qquad (5.14)$$

Therefore, *Newton's gravitational field can be understood as a curvature of the simultaneity sections produced by the matter distribution of density* ρ. Since the Newtonian gravitational potential is static, this interpretation can be extended to the curvature of the Newtonian space–time.
As it happened with the Galilean space–time, Newton's gravitational law, including the above geometric interpretation, is not valid in an arbitrary coordinate system. The coordinate transformations must also involve the gauging or calibration of the Newtonian gravitational potential and vice versa. Therefore, the resulting group of symmetry of Newton's law, called the generalized Galilean group, is given by (exercise)

$$\text{Generalized Galilean group} \quad \begin{cases} x^{i} = \sum A^{i}_{j} x_j + c^{i}(t), \quad i, j = 1, \ldots, 3 \\ t' = at + b, \quad a, b \text{ constants}, a \neq 0 \\ \phi' = \phi - \sum_{i=1}^{3} \dfrac{d^2 c^i}{dt^2} x^i \end{cases} \qquad (5.15)$$

where again A is an orthogonal matrix $AA^T = 1$ and $c^i(t)$ are three differentiable functions of the Newtonian time, whose values depend on how ϕ changes. When $c(t) =$ constant we recover the Galilean group.

5.3 The Minkowski Space–Time

The original equations of the electromagnetic field were written as if they were defined in the Galilean space–time, invariant under the Galilean group. From these equations we derive the wave equations for the electric field **E** and magnetic field **B**, with speed propagation equal to the speed of light in vacuum c. This conveyed the idea that light is an electromagnetic phenomenon which propagates with finite speed in vacuum. Since the Galilean space–time required instantaneous interactions at distance, the propagation of electric and magnetic waves in the Galilean space–time required the existence of a hypothetical media, a fluid called the "lumiferous ether" or simply the "ether." However, the negative result of the Michelson and Morley experiment to measure the interference of that "ether" on the propagation of light, repeated several times between 1891 and 1898, showed that the existence of such fluid was not consistent with the description of light as an electromagnetic phenomenon. In spite of several alternative explanations, in 1904 Lorentz showed that Maxwell's equations are not invariant under the Galilean group, but rather under the *Lorentz/Poincaré group* composed of pseudo-rotations described by a pseudo-orthogonal matrix A and translations p^α:

$$x'^\alpha = A^\alpha_\beta x^\beta + p^\alpha$$

where A is such that $\mathbf{A}\eta\mathbf{A}^T = \eta$, with

$$\eta = \begin{pmatrix} 1 & & & \\ & 1 & & \\ & & 1 & \\ & & & -1/c^2 \end{pmatrix} \tag{5.16}$$

and where c denotes the speed of light in vacuum. The Galilean transformations are obtained in the particular case where $v << c$.

This convinced Einstein in 1905 that the ether did not exist and therefore that the Galilean group should be replaced by the Lorentz/Poincaré group. It was also necessary to modify the traditional law of addition of velocities $v = v_1 + v_2$ based in the instantaneous signal propagation to a new expression compatible with finite speed for light:

$$v = \frac{v_1 + v_2}{1 + \frac{v_1 v_2}{c^2}}$$

When $v_1, v_2 << c$ we obtain the Galilean law of addition of velocities. If v_1 is the velocity of the source and v_2 is the light speed c, the above addition law gives again c, showing that light speed in vacuum is constant in value and also in its independency from the speed of the source.

However, it was only in 1908 that Hermann Minkowski proposed a new geometry compatible with the Lorentz/Poincaré transformations and with Einstein's special relativity, inventing the Minkowski space–time in his opening speech. This is too well known, but it is never too much to remind that this is an experimental result as it was emphatically said by Minkowski: *"The views of space and time which I wish to lay before you have sprung from the soil of experimental physics, and therein lies their strength. They are radical. Henceforth space by itself, and time by itself, are doomed to fade away into mere shadows, and only a kind of union of the two will preserve an independent reality* [73]."

Of course, there is no room for the absolute time in the Minkowski conception of space–time, but the concept of time is still present as a local time, which is nothing but the arc-length of a time-like curve, with the physical dimension adjusted to time: $d\tau = -ds/c$.

Contrasting with the Galilean and the Newtonian space–times which have affine geometries, the Minkowski space–time $\mathscr{M}_4$ has a *metric geometry* defined by the metric (5.16) written in Cartesian coordinates. Thus, the norm of a vector in the Minkowski space–time is

$$\|w\|^2 = \eta_{\mu\nu} w^\mu w^\nu = (w^1)^2 + (w^2)^2 + (w^3)^2 - c^2 (w^4)^2$$

where the minus sign in the fourth component implies that we may have a *null vector*, with zero norm but not being identically zero. When such null vector denotes the position of a point (an event) in Minkowski's space–time, we obtain the equation of a cone

$$(x^1)^2 + (x^2)^2 + (x^3)^2 - c^2 (x^4)^2 = 0$$

The existence of null vectors requires the adaptation of some well-established theorems that are traditionally proved only for Euclidean metrics.[1] In geometry we talk about a "null curve" when its tangent vector is a null vector. An infinitesimal arc element of such a curve is (denoting $dx^4 = d\tau$)

$$ds^2 = (dx^1)^2 + (dx^2)^2 + (dx^3)^2 - c^2 d\tau^2 = 0$$

Denoting the three-dimensional components of the local velocity vector by $v_i = dx^i/d\tau$, we obtain

$$v_1^2 + v_2^2 + v_3^2 = c^2$$

so that the "light cone" is a three-dimensional surface of $\mathscr{M}_4$ where particles travel with the speed of light.

Since the Poincaré transformations do not treat time as a separate parameter, the absolute Newtonian time no longer exists in $\mathscr{M}_4$. In its place, each point (or better, each event) has its own "proper time," as a true local coordinate. With the elimination of the absolute time, the Newtonian concepts of instantaneous communications at distance and of simultaneity sections no longer exist.

[1] For some reason these revisions are frequently referred to as "analysis on Lorentzian metrics" in mathematical texts.

In 1909–1910 it was found that Maxwell's equations are also invariant under a larger 15-parameter group, called the conformal group denoted by C_0. It contains the 10-parameter Poincaré group as a subgroup, plus a four-parameter special transformation, plus the one-parameter metric dilation $\eta_{\mu\nu} = \Omega\eta_{\mu\nu}$, and finally a parameterless inversion $|X|' = 1/|X|$ taking the origin $X = 0$ to infinity [74, 75].

The conformal symmetry was soon forgotten because it requires that the electromagnetic wave equation maintained the advanced potential $A(x + vt)$ as well as the retarded potential $A(x - vt)$. We normally keep only the retarded potential simply because the advanced component appears before the wave is emitted.

Causality was at that time a stronghold of physics: *"The past must be divided from the future by the present. A denial of these facts would be a denial of our most primitive intuitions about time-order"* [76]. This requirement has been recently neglected, perhaps because as it happens with space, time is also more complicated. To understand this we need to discuss in the next concept of space–time.

5.4 Space–Times in General Relativity

In the same way as the Newtonian space–time generalizes the Galilean space–time by incorporating the gravitational field in its geometry, we may say that general relativity generalizes the Minkowski space–time by incorporating the relativistic gravitational field in its geometry.

It is not sufficient to adapt the Newtonian gravitational field to Minkowski's space–time because of the difference of symmetries between the two theories. In fact, denoting by U^μ the components of the velocity vector of a particle in the Minkowski space–time, the right-hand side of Poisson's equation (5.5) can be written as

$$4\pi\rho = \frac{4\pi}{c^2}\rho\eta_{\mu\nu}U^\mu U^\nu$$

The energy–momentum tensor of a system of non-interacting particles (a dust) in special relativity is

$$T^{\mu\nu} = \rho U^\mu U^\nu$$

Replacing these expressions in (5.5), we obtain

$$\eta_{\mu\nu}\frac{\partial^2\phi}{\partial x^\mu \partial x^\nu} = -\frac{4\pi}{c^2}T_{\mu\nu}$$

or, using the fact that the Newtonian gravitational potential is static (do not depend on time), the equation of the Newtonian gravitational field written in the Minkowski space–time reads as

$$\square^2 \phi = -\frac{4\pi}{c^2} T_{\mu\nu} \tag{5.17}$$

where $\square^2$ denotes the D'Alambertian operator $\eta_{\mu\nu}(\partial^2/\partial x^\mu \partial x^\nu)$. However, this is just a cosmetic effect which does not change the fact that (5.5) is invariant under the Galilean group, including the existence of the absolute time, and not under the Lorentz/Poincaré group. A new equation which is consistent with special relativity is required.

As in the Newtonian case, the new gravitational equation can be derived starting from the principle that relativistic free falling particles also follow a geodesic. However, we no longer have the simultaneity sections and light signals propagate along null geodesics. With these observations we may derive a similar geodesic deviation equation (5.12).

Again using a co-moving coordinate system where T has the direction of e_4 and using the metricity condition $g_{\mu\nu;\alpha} = 0$ required in general relativity, we obtain an equation similar to (5.13) except that now the indices count from one to four and we are not required to take the trace of the equation

$$R^\beta_\alpha = -a^\beta{}_{;\alpha} \tag{5.18}$$

From this point on, the arguments are different from the Newtonian case, because instead of a system of free particles Einstein's theory admits a more general distribution of particles, which may or may not interact with each other, each one following a trajectory α and velocity U^α, described by a symmetric energy–momentum tensor $T_{\alpha\beta}$. Therefore, the above relativistic equations (5.18) can be expressed in terms of the symmetric tensor $T_{\alpha\beta}$ as

$$R_{\mu\nu} = -8\pi G T_{\mu\nu}$$

where the factor 8 appears in the right-hand side because $T_{\mu\nu}$ is a symmetric tensor (so that instead of $4\pi\rho$ as in (5.14) we could write $4\pi(T_{\mu\nu} + T_{\nu\mu})$, which is the same as $8\pi T_{\mu\nu}$).

Einstein noted that the above equation is inconsistent, because the energy–momentum tensor is always conserved in the sense that $T^{\mu\nu}{}_{;\nu} = 0$ (we will detail this together with Noether's theorem in Chapter 6), but the left-hand side of the above equation does not satisfy the same condition. To correct this problem Einstein replaced the Ricci tensor $R_{\mu\nu}$ by another tensor

$$G_{\mu\nu} = R_{\mu\nu} - \frac{1}{2} R g_{\mu\nu}$$

called the *Einstein tensor*, constructed with $g_{\mu\nu}$ and its derivatives up to second order and such that $G^{\mu\nu}{}_{;\nu} = 0$ (the contracted Bianchi identity). Cartan showed that the most general solution for this condition is

$$G_{\mu\nu} = R_{\mu\nu} - \frac{1}{2}Rg_{\mu\nu} + \Lambda g_{\mu\nu}$$

where Λ is a constant, called the cosmological constant. Thus, replacing $R_{\mu\nu}$ by this tensor in the above equation we obtain Einstein's equations

$$G_{\mu\nu} = 8\pi G T_{\mu\nu} \tag{5.19}$$

The energy–momentum tensor of a perfect fluid with density ρ and pressure p is written in co-moving coordinates as

$$T_{\mu\nu} = (p+\rho)U_\mu U_\nu - pg_{\mu\nu}$$

The inclusion of Λ in (5.19) has been controversial, starting with Einstein himself. It was not included in the original equation of 1916. Then, in his search for a cosmological solution he proposed the inclusion of Λ. Soon after, he withdrew the constant saying that it was a mistake. In 1917 William deSitter showed to Einstein cosmological solutions generated only by $\pm\Lambda$, called today the deSitter and anti-deSitter solutions, respectively.

Since Λ arises from the contracted Bianchi identity, it can be determined when the appropriate boundary conditions are considered. Using the present astronomical data, the measured value of Λ is $(10^{-46}\, Gev^4/c^2)$, so that it does not play any significant role in local gravitational fields. However, at the cosmological scale of distances it has been claimed to be the explanation for a phenomenon observed since 1998, called the accelerated expansion of the universe. Indeed, taking the plus sign and placing $\Lambda g_{\mu\nu}$ on the right-hand side of (5.19), it can be interpreted as the energy–momentum tensor of a special fluid with state equation $p = -\rho$. At the cosmological scale, the negative pressure exerts a tension which would explain the accelerating effect on the expansion of the universe.

It is also possible to give a physical interpretation to Λ, by placing it on the right-hand side of (5.19) and comparing it with the quantum fluctuations of the vacuum states (see Chapter 7) in quantum field theory. It is found that the vacuum states sum up to a constant value $< \rho_{vac} >$, so that $-8\pi G < \rho_{vac} > g_{\mu\nu}$ could cancel with $\Lambda g_{\mu\nu}$. However, it was found that this theoretical value is much larger, about 10^{120} orders of magnitude than the observed value. Such enormous difference cannot be explained with the known procedures in quantum field theory. This "cosmological constant problem" is currently regarded as the most difficult problem in theoretical physics [77].

The *principle of general covariance* states that any coordinate system can be used to write Einstein's and all pertinent equations. As we have seen in Chapter 1, the coordinates on a manifold are related by the chart composition

$$\phi = \sigma \circ \tau^{-1} : I\!R^n \to I\!R^n$$

which is a diffeomorphism of $I\!R^n$. Therefore the principle of general covariance is also referred to as *diffeomorphism invariance*. This principle represents a strong contrast with Newtonian theories and special relativity, which specify special types of coordinates.

The *equivalence principle* is another postulate of general relativity, allowing the distinction of a gravitational field produced by a mass from that produced by an accelerated system [78]. It can also be seen as a statement restoring the meaning of physical manifold given to a space–time solution of Einstein's equations. Indeed, from the mathematical definition and the discussion in Chapter 2, a manifold is locally equivalent to $I\!R^n$. On the other hand as a space of perceptions involving observers and observables, the physical manifold only makes sense if we have at least two separated events, and therefore it would never be locally equivalent to $I\!R^4$ unless it is flat.

Einstein's equations are derived from the Einstein–Hilbert variational principle

$$\frac{\delta}{\delta g_{\mu\nu}} \int R\sqrt{-g}dv = 0$$

where $R = g^{\mu\nu} R_{\mu\nu}$ is the Ricci scalar curvature derived from the Riemann tensor. The factor $\sqrt{-g}$ is the determinant of the Jacobian matrix of a coordinate transformation, required in conformity with the diffeomorphism invariance and with the transformation of coordinates in a volume integral. The meaning of the Einstein–Hilbert principle seems clear, although it is seldom mentioned: It means that the space–time based on Riemann's geometry of the curvature is the smoothest possible.

The concept of time in general relativity is that of a mere coordinate. Together with the principle of general covariance, time has lost its special characteristic which it has in Newtonian mechanics and in special relativity.

Einstein's equations admit solutions which are compatible with a topology of the type $I\!R^3 \times I\!R$ regardless of whether the metric is static or not. These solutions constitute the bulk of the physical tests of general relativity. However, when we try to apply these topologically special solutions to situations where a dynamical structure is strictly dependent on time, the theory fails. The prime example where this occurs is given by the ADM (or 3+1) decomposition of space–time to the canonical quantum gravity program. In quantum mechanics (in the Schrödinger) sense the Hamiltonian operator is equivalent to the time translation operator. As it happens the Hamiltonian vanishes in a covariant formulation of the ADM decomposition. Several attempts have been made to fix this problem, as for example using Dirac's procedure for constrained systems. It failed to work because the Poisson bracket structure is not diffeomorphism invariant. This is called the time problem in general relativity, and it remains open [79]. One possible alternative is to consider the symmetry of the space–time foliation [80].

Perhaps the ultimate meaning of the principle of general covariance is that time is not relevant to the description of the space–time manifold, except in particular situations. After all, it is possible to describe motion by means of any parameter. Indeed, given a point p in a physical manifold and a tangent vector v_p, we may construct a continuous curve with arbitrary parameter y, defined by a map $\alpha(y) = h_y(p)$, so that for any value of y, regardless of whether it is time like or space like, it can even be a scalar field defined on Euclidean space–time, we obtain a continuous sequence of points in the same physical space. This continuous sequence of points does not follow from any postulated dynamics, but from a well-known group structure, called the *one-parameter group of diffeomorphism* given by the composition map $h_y o h'_y(p) = h_{y+y'}(p)$, $h_0(p) = p$, $h_{y-y}(p) = h_0(p)$. The curve $\alpha(y)$ is called the orbit of p and its tangent vector is the velocity of propagation. Now, given any observable Ω, we may evaluate its variation along the orbit, by the Lie derivative $\pounds_{\alpha'}\Omega$ [81].

Chapter 6
Scalar Fields

Classical field theory results from a natural extension of classical mechanics, when the system of particles is extended to the limit where their coordinates are no longer enumerable. A simple but very intuitive example is given in [55, 82].

We emphasize what we have already said in Chapter 1: Point particles do not exist as a physically observable object, carrying mass, charge, momentum, spin, and other attributes in space–time. Regardless of these attributes they always correspond to *mathematical points* in the parameter space $I\!R^n$ which is not locally equivalent to the physical manifold.

By the duality principle of quantum mechanics all fields satisfying a wave equation are related to a particle with a certain spin or intrinsic angular momentum. The spin-statistics theorem says that particles with integer spin obey the Bose–Einstein statistical interpretation (so they are called *bosons*). Thus, scalar fields correspond to particles of spin 0. Vector fields correspond to particles of spin 1 and symmetric tensor fields of second order correspond to particles of spin 2. On the other hand, particles of half-integer spin are described by spinor fields and they obey the Fermi–Dirac statistical interpretation and for this reason they are called *fermions*.

Scalar fields play an important role in field theory, mainly because the simplicity of its structure (there is just one component) and their derivatives originate the force fields. The best known scalar field is the Newtonian gravitational field or scalar potential, and perhaps the most sought after scalar field (or fields) is the Higgs field (which has not yet been seen in nature), but its existence is required for the success of the whole field theory program. We shall see why it is so in the next sections.

Unless explicitly stated we will be working in Minkowski space–time $\mathscr{M}_4$. As already mentioned all fields, including the scalar fields (seen as a tensor of order $[{}^0_0]$), are defined in a vector bundle based on Minkowski's space–time and with total space $T\mathscr{V}$

$$(\mathscr{M}_4,\ \pi,\ T\mathscr{V})$$

Therefore, in a very general sense we define the following.

M.D. Maia, *Geometry of the Fundamental Interactions*,
DOI 10.1007/978-1-4419-8273-5_6, © Springer Science+Business Media, LLC 2011

Definition 6.1 (Physical Field) A physical field on a manifold $\mathcal{M}$ is a map

$$\Psi : \mathcal{M} \rightarrow T\mathcal{V}$$

such that to each point $p \in \mathcal{M}$, it associates a vector belonging to a local vector space $\mathcal{V}_p$. Furthermore, the field is a solution of the Euler–Lagrange equations defined by a Lagrangian functional, resulting from the action principle

$$\delta A = \delta \int_{\Omega} \mathcal{L}(\Psi, \Psi_{,\mu}, x^{\mu})\, d^4x = 0$$

where Ω is a region in space–time with boundary $\partial\Omega$.

The necessity and existence of an *action principle* in the above definition is not present in the mathematical definition of a vector field, but it is unavoidable in physics.

The idea of an action principle originated with Pierre de Maupertuis when he was the director of the Berlin Academy of Science, when searching for a mathematical proof for the existence of God [83]. His reasoning was *if everything in nature was created by God, then a mathematical function capable of describing the properties of all object in nature would be a mathematical representation of God.*

Maupertuis went as far as proposing the mathematical expression, called the ("divine") *action* as

$$A = \text{matter} \times \text{motion} \times \text{space}$$

Obviously this action is clearly too simple to accomplish such grand objective, and it triggered heavy criticisms by contemporary German mathematicians. In his defense, his friend Leonhard Euler proposed a generalization of the Maupertuis action, stating that it should be an integral over the sum of *elementary actions* of the Maupertuis type $dA = mvds$. This sum led to the Euler *action integral*

$$A = \int mvds$$

where m represents the mass (matter), with velocity v (motion), and ds is the distance (space) [84]. Euler also implemented the condition that the actual description of nature resulted from the extremal values of the action (a maximum or a minimum). In other words, the condition to reproduce a physical effect is that the action should be *stationary*: $\delta A = 0$.

The Euler formulation is considerably better than the Maupertuis action, but clearly it is not sufficient to describe nature. Thus, Lagrange suggested the idea of integrating infinitesimal actions of more elementary processes, defined by a general function of the field variables and its first derivative (a functional), composing a physical system. This function is what we call today the Lagrangian of the physical system [85].

The proposal of Lagrange provides the most general action for the Maupertuis proposition, but without his theological motivation. Yet, the fact that the action principle works remains somewhat mysterious: For every known fundamental theory there is a Lagrangian from which we derive the equations describing the dynamics of the system. Furthermore, as we shall see in Chapter 8, from the Lagrangian and its symmetries we may derive the observables of the theory. We are not very far from Maupertuis idea when looking for a unified theory of fields.

An improvement over such metaphysical situation was achieved by William Rowan Hamilton in 1833, when he modified the action principle, replacing the Lagrangian by a new function, called today the Hamiltonian, which has the interpretation of energy. We shall return to this point below.

Denoting a general field by Ψ the Lagrangian is a functional depending on the field and its first derivatives[1] seen as independent variables. Thus the Euler–Lagrange action writes as

$$A = \int_{\Omega} \mathscr{L}(\Psi, \Psi_{,\mu}) dv$$

where the integration extends over the volume of the region Ω of space–time where Ψ is defined. Under the condition that the variation $\delta\Psi$ vanishes in the boundary $\partial\Omega$, the variational principle $\delta A = 0$ gives the Euler–Lagrange field equations [55].

$$\frac{\partial \mathscr{L}}{\partial \Psi} - \frac{\partial}{\partial x^{\mu}}\left(\frac{\partial \mathscr{L}}{\partial \Psi_{,\mu}}\right) = 0 \tag{6.1}$$

describing the dynamics of field Ψ. Since any field is defined as $\Psi : \mathscr{M} \to T\mathscr{V}$, they are functions of the coordinates. Their partial derivatives $\Psi_{,\mu}$ are often referred to as the "field velocity" in analogy with the case of classical particle physics. Thus, the field Lagrangian is a function defined on the total space $T\mathscr{V}$, acting as the configuration space of the field $\mathscr{L} : T\mathscr{V} \to I\!R^3$ and as such is often called a functional in the sense of a function of functions of the coordinates.

Complementing the analogy with the particle mechanics we may define the "*field momentum*" by the components

$$\pi_{\mu} = \frac{\partial \mathscr{L}}{\partial \Psi_{,\mu}}$$

Usually $\mathscr{L}$ is a quadratic function of the field velocities $\Psi_{,\mu}$, and from its definition the momentum turns out to be a real linear function of the field velocities. In this sense, the momentum can be seen as a linear form on $T\mathscr{V}$, which as we have seen belong to the dual space $T\mathscr{V}^*$. This space is isomorphic to $T\mathscr{V}$, where the isomorphism is non-natural, depending on the definition of a basis of the field space $T\mathscr{V}$.

[1] Higher derivatives are also considered in special cases.

Using a Legendre transformation we obtain the Hamiltonian of the field defined by a map $\mathscr{H} : T\mathscr{V}^* \to I\!R$

$$\mathscr{H}(\Psi, \pi) = \sum \pi_\mu \Psi_{,\mu} - \mathscr{L}$$

in which the explicit dependence on $\Psi_{,\mu}$ is replaced by π_μ. The action integral derived by Hamilton is

$$A = \int \mathscr{H} dv \tag{6.2}$$

Under the condition that $\delta A = 0$, we obtain the Hamilton equations

$$\frac{d\Psi}{d\tau} = \frac{\delta \mathscr{H}}{\delta \pi} \tag{6.3}$$

$$\frac{d\pi}{d\tau} = -\frac{\delta \mathscr{H}}{\delta \Psi} \tag{6.4}$$

Therefore we have at least two formulations for the classical field theory: the *representation space formulation* using the Lagrangian and the Euler–Lagrange equations and the *phase space formulation* using the *Hamiltonian* and Hamilton's equations [55, 86, 87].

The Hamiltonian formulation is the result of an effort of Hamilton to understand the nature of the variational principles. Why does (6.2) work in the sense that they reproduce the known laws of physics? The explanation given by Hamilton is that for closed physical systems, where the potential energy is conserved, the Hamiltonian is the *total energy of the system* $\mathscr{H} = T + U$. Then, nature as a closed system works in such a way that *it spends the minimal energy*. This is a reasonable explanation in replacement for the theological proposition of Maupertuis, as long as we know about the total energy of the universe. As we recall from the introduction, only about 4% of it is known. In this context Leibniz interpreted Hamilton's principle stating that the universe in which we live is the best among all possible universes.

Since the total energy of a closed system is conserved, it is of particular importance for the cases where the field has the smallest potential energy. Using the language of "state of a system" in quantum mechanics, this is referred to as the *state of minimum energy* of the system or the vacuum state of the system defined by

$$\frac{\partial U}{\partial \Psi} = 0$$

Thus, the vacuum state does not necessarily mean the same as empty (or void) space in the sense of classical mechanics, but a state from which we cannot extract further energy.

6.1 Classic Scalar Fields

Consider a vector bundle $(T\mathcal{V}, \pi, \mathcal{M})$ where the total space $T\mathcal{V} = \mathcal{F}(\mathcal{M})$ is composed of the spaces of real (or complex) differentiable functions defined on $\mathcal{M}$. A scalar field on $\mathcal{M}$ is a map

$$\varphi : \mathcal{M} \to \mathcal{F}(M)$$

which associates with each point in $\mathcal{M}$ a real (or complex) value $\varphi(p)$, satisfying a variational principle defined by the Lagrangian $\mathcal{L}(\phi, \phi, \mu, x^{\mu})$.

Let us review the most common examples:

Example 6.1 (The Newtonian Gravitational Field) Newton's gravitational field is perhaps the most common scalar field that we feel in our everyday life.

Consider again the second Newton's law for gravitation:

$$\mathbf{F} = G\frac{m \times m'}{r^3}\mathbf{r}$$

where *Newton's gravitational constant* G was introduced to convert the physical units $[M]^2/[L]^2$ to the units of force. The gravitational potential satisfies the Poisson equation (5.5)

$$\nabla^2\phi = -4\pi G\rho$$

Notice the important fact that G is a dimensional constant. As we have seen, it depends on the integration over solid angles for spherical shells. Thus, if the dimension of space changes to something else, the solid angle also changes and G must adjust its physical dimensions accordingly.

Newton's gravitational equation has agreed with the observations for a wide range of distances. Its validity for small distances has been tested using the Casimir effect, telling that it holds up to 10^{-3} mm, but showing signs of disagreement in the range below 10^{-4} mm [88].

On the other hand, one of the earliest tests of general relativity shows that Newton's gravitation does not describe correctly the Mercury perihelion, a sign that it may not hold beyond the Solar System. In spite of this, it has been assumed to hold near the core of spiral galaxies, although the motion of stars in these galaxies away from the core does not agree with Newton's law [1, 2].

The Lagrangian for the Newtonian gravitational field is

$$\mathcal{L} = \frac{1}{2}\sum \delta^{ij}\phi_{,i}\phi_{,j} + 4\pi G\rho$$

Due to the absolute character of the Newtonian time, the Euler–Lagrange equations (6.1) must be written with the time variable separated from the space coordinates:

$$\frac{\partial \mathscr{L}}{\partial \phi} - \frac{\partial}{\partial x^i}\left(\frac{\partial \mathscr{L}}{\partial \phi_{,i}}\right) - \frac{\partial}{\partial t}\left(\frac{\partial \mathscr{L}}{\partial \dot{\phi}}\right) = 0$$

where $\dot{\phi}$ denotes the derivative with respect to the absolute time. As we have seen in Chapter 5, this field induces an affine connection in Newton's space–time and Poisson's equation can also be written in terms of the Riemann curvature tensor.

Example 6.2 (The Klein–Gordon Field) The Klein–Gordon field appeared in an attempt to derive the relativistic electron equation in 1926, before the discovery of Dirac's equation [89]. The starting point is Schrödinger's equation in quantum mechanics

$$i\hbar \frac{d\Psi}{dt} = \hat{\mathscr{H}}\Psi \tag{6.5}$$

where $\hat{\mathscr{H}}$ denotes the Hamiltonian operator and t is the absolute time. A coordinate transformation that involves a mixture of time and space coordinates will not keep the same equation. Indeed, consider a system composed of a single free particle with mass m and momentum $\wp$ has a Hamiltonian composed only of the kinetic energy $\mathscr{H} = E = \wp^2/2m$. Applying the classical quantum correspondence $\wp \leftrightarrow \hbar\nabla$, $\mathscr{H} \leftrightarrow \hat{\mathscr{H}}$ (6.5) becomes

$$i\hbar \frac{d\Psi}{dt} = \frac{\hbar^2}{2m}\nabla^2\Psi$$

Suppose that this equation is written in the Minkowski space–time with metric $\eta_{\alpha\beta}$, and invariant under the Lorentz transformation:

$$x'^{\alpha} = \sum A^{\alpha}{}_{\beta} x^{\beta}, \quad A\eta A^T = \eta$$

Denoting the transformed wave function by $\Psi'(x')$, the transformation of the Schrödinger equation gives

$$i\hbar\left(A_4^4 \frac{\partial \Psi'}{\partial x'^4} + \sum A_4^i \frac{d\Psi'}{dx'^i}\right) = \frac{\hbar^2}{2m}\left(\sum \delta^{ij}\frac{\partial^2 \Psi'}{\partial x'^i \partial x'^j} + A_4^{\alpha} A_4^{\beta}\frac{\partial^2 \Psi'}{\partial x'^{\alpha}\partial x'^{\beta}}\right)$$

where we have denoted $x^4 = ct$ and $x'^4 = ct'$.

This requires the particular Lorentz transformation that maps time into time only, that is, when $A_4^4 = 1$ and $A_4^i = 0$. However, the first term (the Laplace operator) on the right-hand side is not invariant under a more general Lorentz transformation. The second term is not invariant at all and it cannot be eliminated, except if we impose an additional condition on Ψ'. Therefore, Schrödinger's equation (6.5) is indeed non-relativistic.

One possible explanation for the above result is that we have used the wrong expression for the kinetic energy $\wp^2/2m$, which is Newtonian and distinct from the relativistic kinetic energy which is

$$E^2 = -\wp^2 c^2 + m_0^2 c^4 \tag{6.6}$$

where m_0 denotes the rest mass and $\wp = (p_1, p_2, p_3)$. Replacing this expression in (6.5), we obtain

$$i\hbar \frac{d\Psi}{dt} = \sqrt{-\wp^2 c^2 + m_0^2 c^4}\ \Psi$$

Again, applying the correspondence principle $\wp \to i\hbar\nabla$ and after expanding the root term, the above expression leads to the equation

$$i\hbar \frac{d\Psi}{dt} = m_0 c^2 \sqrt{1 - \frac{\hbar^2}{m_0 c^2}}\Psi \approx \left(1 - \frac{\hbar^2 \nabla^2}{m_0 c^2} + \frac{\hbar^4 \nabla^4}{2 m_0^2 c^4} \cdots \right)\Psi$$

Neglecting the higher order powers of $\hbar$ in this expansion, we obtain again something proportional to $\nabla^2 \Psi'$ and, therefore, even using the relativistic kinetic energy (6.5) is not invariant under the Lorentz symmetry.

The above failure to write a relativistic Schrödinger equation could be explained as a consequence of the series expansion. This could be avoided by taking *the square of the operators* in Schrödinger's equation. But, of course, the result would be an entirely new equation. Indeed, replacing $d\Psi/dt$ by $d^2\Psi'/dt^2$, we obtain the equation describing a new field denoted by φ

$$-\hbar^2 \frac{d^2\varphi}{dt^2} = E^2 \varphi$$

where E is the relativistic kinetic energy (6.6):

$$c^2 \hbar^2 \left(\frac{\partial^2}{c^2 \partial t^2} - \nabla^2 - \frac{m_0 c^2}{\hbar^2} \right) \varphi = 0$$

Denoting $x^4 = ct$ and $m^2 = m_0 c^2/\hbar^2$ we obtain the Klein–Gordon field describing a relativistic field φ equation

$$\left(\Box^2 - m_0^2\right)\varphi = 0 \tag{6.7}$$

which is truly invariant under the Lorentz transformations.

Notice that in principle φ can be any type of field but when it is a scalar function it correctly describes a particle with mass m and spin 0, the neutral scalar meson. However, it does not describe the intended relativistic electron which was its original motivation.

The Klein–Gordon equation can be derived from Lagrangian

$$\mathscr{L} = \frac{1}{2}\left(\eta^{\mu\nu}\varphi_{,\mu}\varphi_{,\nu} + m^2\varphi^2\right) \tag{6.8}$$

where the mass of the spin-0 particle appears as a coefficient of φ^2. The emergence of a term proportional to square of the field in the Lagrangian of an arbitrary field has been consistently associated with mass and therefore it is called the *mass term* of the field. We shall see the importance of this in subsequent examples.

To find the Hamiltonian of the Klein–Gordon field, we start by defining the momentum

$$\pi^\mu = \frac{\partial\mathscr{L}}{\partial\varphi_{,\mu}} = \eta^{\mu\nu}\varphi_{,\nu}$$

Applying in the Legendre transformation we obtain the Hamiltonian of the Klein–Gordon field φ

$$\mathscr{H} = \sum \pi^\mu\varphi_{,\mu} - \mathscr{L} = \frac{1}{2}\left(\pi^\mu\pi_\mu + m^2\varphi^2\right) = T + U$$

where $T(\varphi_{,\mu}) = \pi^\mu\pi_\mu$ is the kinetic term and $U(\varphi) = m\varphi^2$ is the potential energy of the field.

As a simple exercise, show that the vacuum state of the Klein–Gordon field is the point $\varphi = 0$ in the parabola $(U(\varphi), \varphi)$.

Example 6.3 (Complex Scalar Field) The suggestion of Fock–London to make Weyl's theory compatible with quantum theory by using a unitary transformation like

$$\varphi' = \varphi\, e^{i\theta_0} \tag{6.9}$$

may also apply to scalar fields, where the parameter θ may be independent of the space–time coordinates (called global gauge transformations) or not (called local gauge transformations). In the global case it is easy to see that (6.7) is invariant under such transformation.

Such invariance of the Klein–Gordon implies that the Klein–Gordon scalar field is necessarily complex. Denoting by $\mathscr{V}_C$ the vector space of all complex functions, we construct a complex vector bundle $(\mathscr{M}_4,\ \pi,\ T\mathscr{V})$. Then the complex scalar Klein–Gordon field can be formally defined by a map

$$\varphi : \mathscr{M}_4 \to T\mathscr{V}_C$$

such that it assigns a complex vector at each point of $\mathscr{M}$ and such that it must satisfy a variational principle derived by the Lagrangian

$$\mathscr{L} = \eta^{\mu\nu}\varphi_{,\mu}\varphi^*_{,\nu} + m^2\varphi\varphi^* \tag{6.10}$$

which generalizes the Klein–Gordon Lagrangian. Notice that both φ and its complex conjugate φ^* appear in the Lagrangian. The Euler–Lagrange equations calculated with the complex conjugate φ^* and with respect to φ are, respectively,

$$(\square^2 - m^2)\varphi = 0 \quad \text{and} \quad (\square^2 - m^2)\varphi^* = 0$$

which are the complex conjugates of each other. The Lagrangian (6.10) is invariant under the global gauge transformations (6.9).

On the other hand, remembering the original problem of Weyl, and the solution presented by Fock–London as described in Chapter 1, we may ask if the complex Lagrangian is also invariant under *local gauge* symmetries. To see this consider the local gauge transformation

$$\varphi' = \varphi e^{i\theta(x)} \tag{6.11}$$

Replacing in the Lagrangian (6.10), we obtain

$$\mathscr{L} = \eta^{\mu\nu}\left(\theta_{,\mu}\theta_{,\nu}\varphi\varphi^* + i\theta_{,\mu}\varphi^*_{,\nu}\varphi + i\theta_{,\nu}\varphi_{,\nu}\varphi^*\right) - m^2\varphi\varphi^*$$

We clearly see that the resulting Lagrangian is different from (6.10), meaning that (6.11) is not a symmetry of the complex scalar field.

On the other hand, as we have seen in Chapter 3, a Lie group of transformations such as (6.11) can be completely determined by a sequence of infinitesimal transformations. Therefore, considering only an infinitesimal transformation characterized by small values of its parameters $\theta^2 << \theta$ in (6.11)

$$\varphi' \approx (1 + i\theta)\varphi \tag{6.12}$$

Neglecting higher powers of θ and replacing in $\mathscr{L}(\varphi')$ we obtain

$$\begin{aligned}\mathscr{L}(\varphi') &= \eta^{\mu\nu}\partial_\mu\varphi'\partial_\nu\varphi'^* - m^2\varphi'\varphi'^* \\ &= \mathscr{L}(\varphi) + i\eta^{\mu\nu}\theta_{,\mu}(\varphi\varphi^*_{,\nu} - \varphi^*\varphi_{,\nu}) + 0(\theta^2)\end{aligned}$$

This Lagrangian could be re-written as

$$\mathscr{L}(\varphi) = \eta^{\mu\nu}(\varphi_{,\mu} + i\theta_{,\mu}\varphi)(\varphi_{,\nu} + i\theta_{,\nu}\varphi)^* + m^2\varphi\varphi^*$$

which suggests the definition of a *covariant derivative operator*

$$D_\mu\varphi \stackrel{\text{def}}{=} \varphi_{,\mu} + i\theta_{,\mu}\varphi = (\partial_\mu + i\theta_{,\mu})\varphi \tag{6.13}$$

With this definition the Lagrangian can be written as

$$\mathscr{L}(\varphi) = \eta^{\mu\nu}(D_\mu\varphi)(D_\nu\varphi)^* + m^2\varphi\varphi^* \tag{6.14}$$

Except by the exchange of the partial derivative ∂_μ by the "*covariant derivative*" D_μ, the new form of the Lagrangian looks exactly as (6.10). In other words, if we had written from the start (6.10) with D_μ, as in (6.14), then the scalar field would also be invariant under the local transformations.

In the particular case when θ=constant, the new derivative D_μ becomes the usual partial derivative. As we shall see, the whole concept of gauge theory demands a more complete mathematical analysis based on covariant derivatives. More importantly, we will see also that the existence of such covariant analysis is founded on observations.

6.2 Non-linear Scalar Fields

Here we generalize the Klein–Gordon equations by replacing the mass term $m\varphi$ by an analytical function of φ, $F(\varphi)$

$$\Box^2\varphi + F(\varphi) = 0 \tag{6.15}$$

By analytic we mean that F can be represented by a converging positive power series of φ:

$$F(\varphi) = \sum_0^\infty \lambda_n \varphi^n$$

In particular, assuming that for a given integer N we have $\varphi^{N+1} << \varphi^N$, then the series reduces to a polynomial of degree N. We see that the example of the Klein–Gordon equation is the particular case when $N = 1$ and $\lambda_1 = m^2$. Other choices of $F(\varphi)$ give some very interesting non-linear scalar fields.

Example 6.4 (Higgs Quartic Potential) Consider now that $F(\varphi)$ is a particular polynomial of order 3 with coefficients

$$\lambda_0 = 0,\ \lambda_1 = \mu^2,\ \lambda_2 = 0,\ \lambda_3 = \frac{\lambda}{3!},\ \lambda > 0$$

From (8.8) the field equation becomes

$$(\Box^2 + \mu^2)\varphi + \frac{\lambda}{3!}\varphi^3 = 0$$

which is obtained from the Lagrangian

$$\mathscr{L} = \frac{1}{2}\eta^{\mu\nu}\varphi_{,\mu}\varphi_{,\nu} - U(\varphi)$$

where we have denoted the potential

$$U(\varphi) = -\frac{1}{2}\mu^2\varphi^2 - \frac{\lambda}{4!}\varphi^4$$

Because of the existence of the last term of fourth power in φ, this example is often referred to as the *quartic potential theory* or simply as the $\lambda\phi^4$ *theory*.

Since the equation is non-linear, the field φ is self-interacting, meaning that it can be generated by itself, even in the case of the vacuum.

The above equation can be easily extended to the case of a complex scalar field by replacing $\varphi^2 \to \varphi\varphi^*$. Like in the previous example, the Lagrangian becomes invariant under the local gauge transformations, $\varphi' = e^{i\theta(x)}\varphi$, provided the partial derivatives are replaced by the covariant derivatives $D_\mu = \partial_\mu + i\theta_\mu$ (exercise).

Like in the case of the Klein–Gordon Lagrangian, the vacuum states of this field are solutions of

$$\frac{\partial U}{\partial \varphi} = 0$$

or, equivalently,

$$\varphi\left(\mu^2 + \frac{\lambda}{3!}\varphi^2\right) = 0$$

whose solutions depend on the sign of μ^2. Therefore we have two cases to consider:

(a) If $\mu^2 > 0$ the only solution is $\varphi = \varphi_0 = 0$, where $U(\varphi_0) = 0$.
In this case, the curve (U, φ) shows a parabola with minimum value at zero.

(b) If $\mu^2 < 0$, then besides $\varphi = 0$ we have also two other real solutions

$$\varphi = \sqrt{\frac{-6\mu^2}{\lambda}} = a, \qquad \varphi = -\sqrt{\frac{-6\mu^2}{\lambda}} = -a$$

The curve $(U(\varphi), \varphi)$ is a quartic curve (Fig. 6.1) intersecting a straight line in four points at most.

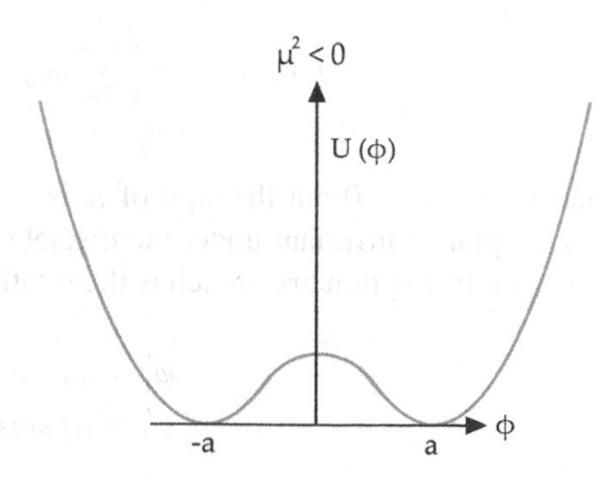

Fig. 6.1 Quartic potential

The vacuum states have a reflection symmetry $\varphi' = -\varphi$, which is discrete in the sense that its parameter assumes only the values 1 and -1. Since the vacuum states of φ are the states of minimum energy, these two vacuum points are the possible points of rest of the field. When the field φ chooses to rest in any of these vacuum states, say $\varphi = a$, then this becomes the state of lowest energy, or the true vacuum. Since they have the same minimal energy they are equivalent and there is no way we can decide which one the field chooses, except by an outside interference.

This is the beginning of an interesting story: Supposing the state of lowest energy has been chosen by the system to be at $\varphi = a$, the number $-\mu^2 = m$ must be real. Comparing with the mass term of the Klein–Gordon Lagrangian, we obtain the equivalent to a "mass term" of the quartic potential theory, as a real factor of φ^2:

$$\mathscr{L}(\varphi) = \frac{1}{2}\left(\eta^{\mu\nu}\partial_\mu\varphi'\partial_\nu\varphi + m\varphi^2\right) - \frac{\lambda}{4!}(\varphi + a)^4$$

We say that the choice of one vacuum state represents the breaking of the vacuum symmetry $\varphi \to -\varphi$. That is, when the symmetry of the vacuum ceases to exist by a spontaneous choice of the vacuum state, we obtain a mass term. This is the basic example of the so-called Higgs mechanism, a new concept in field theory called the *spontaneous symmetry breaking* of the vacuum symmetry introduced in 1964 by P. Higgs and T. B. Kibble [38, 39]. Its importance lies in the obtention of mass by symmetry breaking of a field theory without mass term. The term spontaneous results from the fact that it occurs with minimal energy, independently of any external action.

Example 6.5 (A Pair of Scalar Fields) To obtain further insight into the symmetry breaking mechanism consider the Lagrangian with quartic interaction, involving two independent complex scalar fields:

$$\mathscr{L}(\varphi_1, \varphi_2) = \frac{1}{2}\sum_{i=1}^{2}\eta^{\mu\nu}\varphi^*_{i,\mu}\varphi_{i,\nu} + U(\varphi_i)$$

where the potential energy is

$$U(\varphi_i) = \mu^2\left(\sum_{i=1}^{2}\varphi_i\varphi^*\right) + \frac{\lambda}{4!}\left(\sum_{i=1}^{2}\varphi_i\varphi_i^*\right)^2$$

and where $\lambda > 0$ but the sign of μ is arbitrary. As in the previous example, this Lagrangian is invariant under the discrete symmetry $\varphi_i' = -\varphi_i$. However, here we have another symmetry, which is the rotation in the plane (φ_1, φ_2) given by

$$\varphi_1' = \varphi_1\cos\theta - \varphi_2\sin\theta$$
$$\varphi_2' = \varphi_1\sin\theta + \varphi_2\cos\theta$$

where θ is the rotation angle. Again, this angle can be independent of the coordinates (a global symmetry) or it may depend of the coordinates of the space-time (a local symmetry).

The above Lagrangian can be better understood if we take the two independent scalar fields as generating a *two-dimensional space of functions* with basis $\{\varphi_1, \varphi_2\}$ (since each function is an infinite dimensional vector we may say that this space is a two-dimensional space of functions). In such basis the above transformation may be written in terms of a matrix representation

$$u = \begin{pmatrix} \cos\theta & -\sin\theta \\ \sin\theta & \cos\theta \end{pmatrix}$$

such that $uu^\dagger = 1$ and $\det u = 1$. Therefore we may identify this group as being the unitary group $U(1)$ generated by the single parameter θ. The matrices u of the representation act as $\varphi' = u\varphi$ on the space of columns

$$\varphi = \begin{pmatrix} \varphi_1 \\ \varphi_2 \end{pmatrix}, \quad \varphi^\dagger = (\varphi_1^*, \varphi_2^*)$$

In this notation, notice that $\sum_i \varphi_i^2 = \varphi^\dagger \varphi$ and $\sum_i \eta^{\mu\nu} \varphi^\dagger_{i,\mu} \varphi_{i,\nu} = \eta^{\mu\nu} \varphi^\dagger_{,\mu} \varphi_{,\nu}$.

The vacuum states of this system are given by the solutions of

$$\frac{\partial U}{\partial \varphi_i} = \left(\mu^2 + \frac{\lambda}{3!}(\varphi_1^2 + \varphi_2^2)\right)\varphi_i = 0, \quad i = 1, 2$$

Therefore, for $\mu^2 > 0$ only the origin $\varphi_1 = \varphi_2 = 0$ represents a real vacuum. On the other hand, when $\mu^2 < 0$ (denote $\mu^2 = -m^2$), we obtain an equation

$$\varphi_1^2 + \varphi_2^2 = \frac{6m^2}{\lambda}$$

which defines a circle in the plane (φ_1, φ_2). The infinite points of this circle represent vacuum solutions of the system, which are mapped one onto another by the above-mentioned rotation group.

As in the previous examples, when the system spontaneously chooses to rest in any of these points, we have a spontaneous breaking of the vacuum symmetry. For example, the choice $\varphi_1 = \sqrt{6m^2/a}$ breaks the vacuum symmetry and as in the previous example, we say that φ_1 acquires a mass while φ_2 is a solution that remains without mass (Fig. 6.2).

Next we ask if the same system has a local gauge symmetry, when θ is a function of the coordinates, that is, with the transformation

$$\varphi' = e^{i\theta(x)}\varphi, \quad uu^\dagger = 1$$

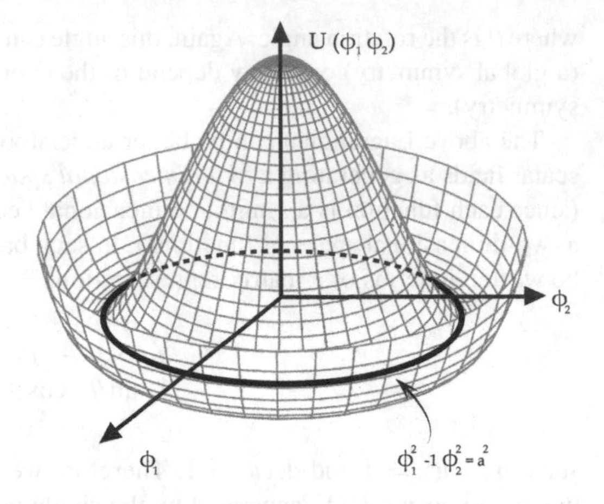

Fig. 6.2 The Mexican hat quartic potential for two scalar fields

In this case the vacuum symmetry is given by the same group $U(1)$, but it is locally defined. Following the same procedure as in the previous examples, the potential energy is not affected but the partial derivatives of the kinetic term must consider the coordinate dependence of the matrices u. Therefore, we have $\varphi'_{,\mu} = u_{,\mu}\varphi + u\varphi_{,\mu}$ and $\varphi'^{\dagger}_{,\mu} = \varphi^{\dagger}u^{\dagger}_{,\mu} + \varphi^{\dagger}_{,\mu}u^{\dagger}$ so that

$$\eta^{\mu\nu}\varphi'^{\dagger}_{,\mu}\varphi'_{,\nu} = \eta^{\mu\nu}\left(\varphi^{\dagger}u^{\dagger}_{,\mu}u_{,\nu}\varphi + \varphi^{\dagger}u_{,\mu}u\varphi_{,\nu} + \varphi^{\dagger}_{,\mu}u_{,\nu}u\varphi\right)$$

As we see, the result is not the same as the original kinetic term $\eta^{\mu\nu}\varphi^{\dagger}_{,\mu}\varphi_{,\nu}$. Therefore the Lagrangian is not invariant under the local $U(1)$ symmetry. However, it is possible to redefine the derivative so as to make the Lagrangian invariant under the local gauge transformations. For that purpose consider an infinitesimal value of θ such that $\theta^2 << \theta$ when we obtain

$$u(\theta) = \begin{pmatrix} \cos\theta & -\sin\theta \\ \sin\theta & \cos\theta \end{pmatrix} \approx \begin{pmatrix} 1 & -\theta \\ \theta & 1 \end{pmatrix} = \sigma_0 - \theta i\sigma_2$$

From this we find that $u\,u^{\dagger} \approx 1$, $u^{\dagger}_{,\mu}u_{,\nu} \approx 0$, and $uu_{,\mu} \approx -i\sigma_2\theta_{,\mu}$. It follows that

$$\eta^{\mu\nu}\varphi'^{\dagger}_{,\mu}\varphi'_{,\nu} = -\eta^{\mu\nu}\left(\varphi^{\dagger}_{,\mu}\theta_{\nu}\varphi + \varphi^{\dagger}\theta_{,\mu}\varphi_{,\nu}\right)i\sigma_2$$

or, after defining the covariant derivative operator

$$D_{\mu} = \sigma_0\partial_{\mu} - \theta_{,\mu}i\sigma_2$$

we obtain the invariant Lagrangian

$$\mathscr{L}(\varphi) = \frac{1}{2}\eta^{\mu\nu}(D_\mu\varphi)^\dagger(D_\nu\varphi) + U(\varphi)$$

The above example may be generalized to the case where we have N scalar fields φ_i. Then we have an important result:

Theorem 6.1 (Goldstone) *Given N independent complex scalar fields, φ_i, $i = 1, \ldots, N$, in a Lagrangian $\mathscr{L} = T - U$, which is invariant under an n-parameter local field symmetry G, with parameters θ_a, $a = 1, \ldots, n$, then there are at most P non-trivial vacuum states, where P is the rank of the matrix $(\partial^2 U(\varphi)/\partial\varphi_i\partial\varphi_j)$.*

Using the same matrix notation, the N scalar fields can be seen as the components of a covariant vector (a column):

$$\varphi = \begin{pmatrix} \varphi_1 \\ \varphi_2 \\ \vdots \\ \varphi_N \end{pmatrix}$$

Then, the Lagrangian of the system can be written as

$$\mathscr{L}(\varphi) = \frac{1}{2}\eta^{\mu\nu}\varphi^\dagger_{,\mu}\varphi_{,\nu} - U(\varphi)$$

Using a basis of the field space comprising N independent vectors $\{\varphi_i\}$, an infinitesimal transformation of the group G acting on that basis is given by

$$\delta\varphi_k = \sum_a \theta^a X_a \varphi_k$$

where X_a denote the operators of the Lie algebra of G. Then, the vacuum states of the system are given by the solutions of

$$\delta_k U(\varphi) = \frac{\partial U}{\partial\varphi_k}\delta\varphi_k = \frac{\partial U}{\partial\varphi_k}\sum_a \theta^a X_a \varphi_k = 0$$

Since all parameters are independent we obtain the condition

$$\frac{\partial U}{\partial\varphi_k}(X_a\varphi_k) = 0$$

Deriving this equation with respect to φ_ℓ, we obtain

$$\frac{\partial^2 U}{\partial\varphi_k\partial\varphi_\ell}X_a\varphi_k + \frac{\partial U}{\partial\varphi_k}X_a\delta_{kl} = 0$$

Since the vacuum solutions satisfy the condition $\partial U/\partial\varphi_k = 0$, the last equation reduces to the matrix equation

$$MX_a\varphi = 0$$

where we have denoted the matrix $M = (\partial^2 U/\partial\varphi_i\partial\varphi_j)$ and X_a are matrices defined in representation of the Lie algebra of G. Therefore, the vacuum states are solutions of a homogeneous system.

There are two cases to analyze:

(a) If $\det M \neq 0$, then multiplying the above equation by M^{-1} we obtain $X_a\varphi = 0$. Since the matrices X_a represent a basis of the Lie algebra of G, it follows that $\varphi = 0$. In this case the vacuum states trivially coincide with the origin of the space of solutions.

(b) If $\det M = 0$, then the homogeneous system has a non-trivial solution, although not complete. That is, the number of non-zero solutions is equal to the number P of lines and columns of the largest sub-matrix of M which has a non-zero determinant (the rank of M). In this case, there are $N - P$ solutions which are not determined by the above equation are called the Goldstone bosons of the system [90, 91].

Example 6.6 (Topological Solitons) Returning to equation (6.15), consider that $F(\varphi)$ is represented by a converging power series to the periodic function $\alpha/\beta \sin(\beta\varphi)$, where α, β are arbitrary constants. Proceeding in the usual way we obtain the field equations

$$\Box^2\varphi + \frac{\alpha}{\beta}\sin(\beta\varphi) = 0 \qquad (6.16)$$

This equation follows from the Lagrangian

$$\mathscr{L} = \frac{1}{2}\eta^{\mu\nu}\partial_\mu\varphi\partial_\nu\varphi - U(\varphi)$$

$$U(\varphi) = -\frac{\alpha}{\beta^2}(\cos(\beta\varphi) - 1)$$

Notice that (6.16) is a non-linear equation whose solutions depend on the constants α and β. A well-known example corresponds to $\beta = 1$ and $\alpha = m^2 > 0$, giving the "*Sine–Gordon equation*"

$$\Box^2\varphi + m^2 \sin\varphi = 0$$

This equation does not have exact solutions in arbitrary dimensions [92]. However, there is a solution in two dimensions known as the topological soliton, representing a stable field with finite energy, which propagates without dispersion. Note also that for small values of φ we have $\sin\varphi \approx \varphi$, so that we recover the linear Klein–Gordon equation.

As an exercise, find all vacuum states for solitons.

Chapter 7
Vector, Tensor, and Spinor Fields

7.1 Vector Fields

The prime example of a vector field is the electromagnetic field in Minkowski space–time. It is an essential component of the development of modern physics, including the emergence of relativity and the relevance of the concept of symmetry in physics. Due to the importance of this combination of theoretical and experimental results to the development of gauge theory let us briefly review the basics of the electromagnetic field (for more details, see, e.g., [93]) and some more advanced topics involving its interactions with scalar fields.

7.1.1 The Electromagnetic Field

The systematic observations of electricity is credited to Stephen Grey and François da Fay between 1736 and 1739. However, real progress was possible only after the Leyden (Holland) bottles were made in 1746. Quantitative results started to show up around 1777 with the invention of the torsion balance by Charles Augustin Coulomb, leading to the Coulomb law:

$$\mathbf{F} = K\frac{qq'}{r^2}\frac{\mathbf{r}}{r}, \quad K = constant$$

The systematic study of the electric current was possible only after 1794 with the invention of the electric battery by Allesandro Volta, allowing for the use of the electric current in a controlled way. The electromagnetism appeared around 1819, after Hans Christian Oersted observed that magnetic forces, originally observed only with permanent magnets, could also be *induced* by the presence of varying electric current. The relation between this magnetic field and the derivative of the current with respect to time was discovered by André Marie Ampére in 1819, leading to a differential expression for the Coulomb law.

Ampère imagined the electric current as formed by small cylindrical moving sections with length $d\ell$ and area A of the conductor, with charge density ρ, so that

M.D. Maia, *Geometry of the Fundamental Interactions*,
DOI 10.1007/978-1-4419-8273-5_7, © Springer Science+Business Media, LLC 2011

the charge in each section is $dq = \rho dV = \rho d\ell dA$. Replacing in Coulomb's law he obtained a differential expression for the electric force

$$|d\mathbf{F}| = K\frac{q\rho A d\ell}{r^2}$$

The experimental observation by Michael Faraday that this force induced a magnetic attraction between two close wires led to the concept of magnetic induction and magnetic field flux and eventually to the Faraday law of 1831 stating that, *The variation of the magnetic flux with time induces an electric current on a conductor which is proportional to that variation.*

The final step in this rather complex development was given by Jean-Baptiste Biot and Félix Savart in 1822, obtaining the expression

$$\mathbf{F}_{\text{mag}} = K'\frac{d\ell \wedge d\ell'}{r}, \quad K' = constant$$

where $d\ell$ and $d\ell'$ are the tangent vectors to two small cylindrical sections of two conductors separated by a distance r.

Thus the electric and magnetic fields which were originally thought to be two independent fields become related to each other. However, the differential second-order equations describing these two fields were not quite consistent. The completion of the consistency process was elaborated in 1861 by James Clark Maxwell [94].

Of course, the electric and magnetic field equations were originally written with the absolute time t and consequently with the idea of simultaneity sections Σ_t as in the Galilean space–time. They were expressed in terms of the scalar ϕ and vector potentials $\mathbf{A}$ as

$$\mathbf{B} = \nabla \wedge \mathbf{A}, \qquad \mathbf{E} = -\nabla\phi - \frac{1}{c}\frac{\partial \mathbf{A}}{\partial t} \tag{7.1}$$

From these expressions we obtain immediately two homogeneous equations

$$< \nabla, \mathbf{B} > = 0, \quad \nabla \wedge \mathbf{E} + \frac{1}{c}\frac{\partial \mathbf{B}}{\partial t} = 0 \tag{7.2}$$

The two remaining equations, the Coulomb and Ampère equations, involve electrical charges and current

$$\nabla^2\phi = -4\pi\rho, \quad \nabla^2 A = -\frac{4\pi}{c}\mathbf{J} + \frac{1}{c}\frac{\partial \mathbf{E}}{\partial t}$$

Originally these equations were inconsistent because the Faraday and Ampère equations hold under different conditions. This was fixed by Maxwell, and today they combine in the four Maxwell's equations

$$< \nabla, \mathbf{B} >= 0 \tag{7.3}$$

$$\nabla \wedge \mathbf{E} + \frac{1}{c}\frac{\partial \mathbf{B}}{\partial t} = 0 \tag{7.4}$$

$$< \nabla, \mathbf{E} >= 4\pi\rho \tag{7.5}$$

$$\nabla \wedge \mathbf{B} - \frac{1}{c}\frac{\partial \mathbf{E}}{\partial t} = -4\pi \mathbf{J} \tag{7.6}$$

Only the last two (non-homogeneous) equations are the Euler–Lagrange equations with respect to the vector potential **A** and the scalar potential ϕ in the Lagrangian

$$\mathscr{L} = \frac{< \mathbf{E}, \mathbf{E} > - < \mathbf{B}, \mathbf{B} >}{8\pi} - \rho\phi - < \mathbf{J}, \mathbf{A} > \tag{7.7}$$

This Lagrangian is invariant under a special local transformation of the potential functions given by

$$\mathbf{A}' = \mathbf{A} + \nabla\theta \tag{7.8}$$

$$\phi' = \phi - \frac{1}{c}\frac{\partial \theta}{\partial t} \tag{7.9}$$

where the parameter θ *is a function of the space–time coordinates*.

The invariance of the Lagrangian under the above transformations follows directly from the fact that in (7.1), the expressions of **E** and **B** are invariant under the above transformations. Indeed

$$\mathbf{E}' = -\nabla\phi - \frac{1}{c}\frac{\partial}{\partial t}\nabla\theta - \frac{1}{c}\frac{\partial \mathbf{A}}{\partial t} + \frac{1}{c}\frac{\partial}{\partial t}\nabla\theta = -\nabla\phi - \frac{1}{c}\frac{\partial \mathbf{A}}{\partial t} = \mathbf{E}$$
$$\mathbf{B}' = \nabla \wedge (\mathbf{A} + \nabla\theta) = \nabla \wedge \mathbf{A} = \mathbf{B}$$

Consequently, Maxwell's equations also do not change under the same transformations.

The set of transformations (7.8) and (7.9) constitute a group with respect to the composition

$$\mathbf{A}' = \mathbf{A} + \nabla\theta, \qquad \mathbf{A}'' = \mathbf{A}' + \nabla\theta'$$
$$\phi' = \phi - \frac{1}{c}\frac{\partial \mathbf{A}}{\partial t}, \quad \phi'' = \phi' - \frac{1}{c}\frac{\partial \theta'}{\partial t}$$

which combine into transformations of the same kind

$$\mathbf{A}'' = \mathbf{A} + \nabla(\theta + \theta') = \mathbf{A} + \nabla\theta''$$
$$\phi'' = \phi - \frac{1}{c}\frac{\partial}{\partial t}(\theta + \theta') = \phi - \frac{1}{c}\frac{\partial\theta''}{\partial t}$$

The identity transformation corresponds to θ = constant. Choosing this constant to be zero, the inverse of a transformation corresponds naturally to $-\theta$. We may easily check that the above composition is associative. Since the order of composition does not affect the result we have an Abelian group, where the only parameter θ is a function of the coordinates. This is the *electromagnetic gauge group*. This group is a Lie group with one coordinate-dependent parameter. Therefore, this group is isomorphic to the local group of rotations $SO(2)$ and as we have seen also to the local unitary group $U(1)$.

With the appropriate choice of conditions imposed on θ, we obtain different solutions of Maxwell's equations. The two most common choices are as follows:

(a) The Lorentz gauge

From (7.8) and (7.9) we may write

$$< \nabla, \mathbf{A}' > = < \nabla, \mathbf{A} > + \nabla^2\theta$$
$$\frac{1}{c}\frac{\partial\phi}{\partial t} = \frac{1}{c}\frac{\partial\phi}{\partial t} - \frac{1}{c}\frac{\partial^2\theta}{\partial t^2}$$

so that

$$\left(< \nabla, \mathbf{A}' > + \frac{1}{c}\frac{\partial\phi'}{\partial t}\right) = \left(< \nabla, \mathbf{A} > + \frac{1}{c}\frac{\partial\phi}{\partial t}\right) + \nabla^2\theta - \frac{1}{c^2}\frac{\partial^2\theta}{\partial t^2}$$

Therefore, assuming that θ is such that

$$\nabla^2\theta - \frac{1}{c}\frac{\partial^2\theta}{\partial t^2} = 0 \tag{7.10}$$

it follows that

$$< \nabla, \mathbf{A}' > + \frac{1}{c}\frac{\partial\phi'}{\partial t} = < \nabla, \mathbf{A} > - \frac{1}{c}\frac{\partial\phi}{\partial t} = C$$

where C is a constant. In particular choosing this constant to be $C = 0$, we obtain the Lorentz gauge condition

$$< \nabla, \mathbf{A} > - \frac{1}{c}\frac{\partial\phi}{\partial t} = 0$$

which is compatible with the electromagnetic wave solution of Maxwell's equations.

(b) The Coulomb gauge
Here we consider a more restrictive condition where the scalar potential does not depend on time. Then we obtain

$$< \nabla, \mathbf{A} > = 0$$

Replacing this in Maxwell's equations we obtain

$$\nabla^2 \phi = 4\pi \rho$$

which is Poisson's equation for a charge density $\rho(x, t)$ and whose solution describes the Coulomb potential for the electrostatic field of an isolated particle.

7.1.2 The Maxwell Tensor

The development of the electromagnetic theory in the beginning of the 20th century led to the conclusion that Maxwell's theory written in the Galilean space–time with its simultaneous sections was not compatible with the explanations of the negative results of the *Michelson–Morley experiment* on the propagation of light within the context of the Galilean space–time. This resulted in the theory of special relativity based on the Minkowski space–time previously described. Then Maxwell's equations are more appropriately written in the Minkowski space–time, using the concept of proper time denoted by τ. Correspondingly, the Galilean group was replaced by the Poincaré group and the light speed was assumed to be a fundamental constant of nature.

The electromagnetic field which was written as a pair of vectors $(\mathbf{E}, \mathbf{B})$ can now be more appropriately written as an anti-symmetric rank two tensor field composed of the components of $\mathbf{E}$ and $\mathbf{B}$ given by (7.1):

$$E_i = -\partial_i \phi - \frac{1}{c}\frac{\partial}{\partial \tau} A_i \tag{7.11}$$

$$B_i = \partial_j A_k - \nabla_k A_j \qquad (i, j, k \text{ cyclic} = 1, 2, 3) \tag{7.12}$$

Define the tensor $F = (F_{\mu\nu})$ by its components

$$F_{ij} = \partial_i A_j - \partial_j A_i, \quad F_{ii} = 0$$

$$F_{i4} = \partial_i A_4 - \partial_4 A_i, \quad F_{44} = 0$$

where we have denoted $\partial_4 = \frac{1}{c}\frac{\partial}{\partial \tau}$. Introducing the potential four-vector

$$A = (\mathbf{A}, -\phi) = (A_1, A_2, A_3, A_4)$$

and denoting its individual components by A_μ, the expressions of $F_{\mu\nu}$ can be summarized as

$$F_{\mu\nu} = \partial_\mu A_\nu - \partial_\nu A_\mu, \quad \mu; \nu = 1, \ldots, 4$$

Therefore, comparing with (7.11) and (7.12) we obtain

$$F_{12}=\partial_1 A_2-\partial_2 A_1=B_3, \; F_{13}=\partial_1 A_3-\partial_3 A_1=-B_2, \; F_{23}=\partial_2 A_3-\partial_3 A_2=B_1,$$

$$F_{i4} = \partial_i A_4 - \partial_4 A_i = -\partial_i \phi - \frac{\partial}{\partial \tau} A_i = E_i, \; F_{44} = 0$$

In a shorter notation we may write

$$F_{i4} = E_i, \qquad F_{ij} = \varepsilon_{ijk} B_k \tag{7.13}$$

where ε_{ijk} is the standard Levi-Civita permutation symbol for $i, j, k = 1, \ldots, 3$.

Explicitly, we obtain an array (it is not a matrix)

$$(F_{\mu\nu}) = \begin{pmatrix} 0 & -E_3 & -E_2 & -E_3 \\ E_1 & 0 & B_3 & B_2 \\ E_2 & -B_3 & 0 & -B_1 \\ E_3 & -B_2 & B_1 & 0 \end{pmatrix}$$

which is known as the covariant Maxwell tensor. The corresponding contravariant tensor has components

$$F^{\mu\nu} = \eta^{\mu\rho} \eta^{\nu\beta} F_{\rho\beta}$$

In order to write Maxwell's equations in terms of the Maxwell tensor, we start with the components of the two non-homogeneous equations (7.5) and (7.6)

$$\sum \partial_i E_i = 4\pi\rho$$

$$\sum \varepsilon_{ijk} \partial_j B_k - \frac{1}{c} \frac{\partial E_i}{\partial \tau} = -4\pi J_i$$

Using (7.13), the two non-homogeneous equations correspond to

$$\sum \partial_i F_{i4} = 4\pi\rho$$

$$\sum \partial_j F_{ij} - \frac{1}{c} \frac{\partial F_{i4}}{\partial \tau} = -4\pi J_i$$

However, $F^{j4} = -F_{j4}$ and $F^{ij} = F_{ij}$ so that

$$-\sum \partial_j F^{j4} = 4\pi\rho \quad \text{(Coulomb)}$$

$$\sum \partial_j F^{ij} + \partial_4 F^{j4} = J \quad \text{(Ampère)}$$

Defining the four-dimensional current $(\mathbf{J}, -c\rho) = (J_1, J_2, J_3, J_4)$ with components J_μ and $J^\mu = \eta^{\mu\rho} J_\rho$, it follows that the above two equations can be summarized as

$$F^{\mu\nu}{}_{,\nu} = \frac{4\pi}{c} J^\mu$$

On the other hand, the two homogeneous Maxwell's equations are

$$\sum \partial_i B_i = 0, \quad \sum \varepsilon_{ijk} \partial_j E_k + \frac{1}{c} \frac{\partial B_i}{\partial \tau} = 0$$

which can also be written in terms of $F_{\mu\nu}$ as

$$\partial_i \varepsilon_{ijk} F_{jk} = 0$$

$$\varepsilon_{ijk} \partial_j F_{k4} + \frac{1}{c} \frac{\partial}{\partial \tau} \varepsilon_{ijk} F_{jk} = 0$$

or, using the four-dimensional Levi-Civita permutation symbol

$$\varepsilon^{\mu\nu\rho\sigma} = \begin{cases} 1 & \text{if } \mu\nu\rho\sigma \text{ is an even permutation of 1234} \\ -1 & \text{if } \mu\nu\rho\sigma \text{ is an odd permutation of 1234} \\ 0 & \text{in any other case} \end{cases}$$

The last two equations can be summarized as

$$\varepsilon^{\mu\nu\rho\sigma} \partial_\nu F_{\rho\sigma} = 0$$

Therefore the four Maxwell's equations are equivalent to

$$F^{\mu\nu}{}_{,\nu} = 4\pi J^\mu \tag{7.14}$$

$$\varepsilon^{\mu\nu\rho\sigma} \partial_\nu F_{\rho\sigma} = 0 \tag{7.15}$$

which are known as the manifestly covariant Maxwell's equations, having the same shape in any Lorentz frame in Minkowski's space–time.

7.1.2.1 The Lagrangian of the Electromagnetic Field

The Lagrangian of the electromagnetic field (7.7) can be written in terms of the Maxwell tensor $F_{\mu\nu}$. For that purpose consider the four-vectors J_μ and A_μ defined previously and define the Lorentz-invariant quantity

$$\eta^{\mu\nu} A_\mu J_\nu = < \mathbf{J}, \mathbf{A} > + \rho\phi$$

On the other hand, from the components of $\mathbf{E}$ and $\mathbf{B}$ written in terms of $F_{\mu\nu}$ we obtain

$$< \mathbf{E}, \mathbf{E} > = \sum E_i E_i = -\sum F_{14}F_{14} - \sum F_{24}F_{24} - \sum F_{34}F_{34}$$

and

$$< \mathbf{B}, \mathbf{B} > = \sum B_i B_i = F_{12}F^{12} + F_{13}F^{13} + F_{23}F^{23}$$

Therefore,

$$< \mathbf{E}, \mathbf{E} > - < \mathbf{B}, \mathbf{B} > = \frac{1}{2} F^{\mu\nu} F_{\mu\nu}$$

where the factor $1/2$ was included to compensate for the repeated terms in the sum of the right-hand side. Therefore, the Lagrangian of the electromagnetic field (multiplied by 4π) is

$$\mathscr{L} = \frac{1}{4} F_{\mu\nu} F^{\mu\nu} - 4\pi J^{\mu} A_{\mu} \tag{7.16}$$

To see that this form of the Lagrangian leads directly to the covariant equations let us calculate the variation with respect to A_μ

$$\frac{\partial \mathscr{L}}{\partial A_\mu} = 4\pi J^\mu \text{ and } \frac{\partial \mathscr{L}}{\partial A_{\mu,\nu}} = \frac{1}{2} F^{\rho\sigma} \frac{\partial F_{\rho\sigma}}{\partial A_{\mu,\nu}} = \frac{1}{2} \left(\delta^{\alpha}_{\rho} \delta^{\nu}_{\sigma} - \delta^{\nu}_{\rho} \delta^{\mu}_{\sigma} \right) F^{\rho\sigma} = F^{\mu\nu}$$

Consequently, the electromagnetic Euler–Lagrange equations are

$$F^{\mu\nu}{}_{,\nu} = 4\pi J^{\mu} \tag{7.17}$$

The other two equations (the homogeneous equations) are obtained directly from the expressions of **E** and **B** in terms of A and ϕ. For this reason they are often referred to as non-dynamical equations. Indeed, they are part of an identity satisfied by $F_{\mu\nu}$ as we shall see later.

7.1.3 The Nielsen–Olesen Model

The Nielsen–Olesen model arose originally from an attempt to describe a quantized magnetic flux [95]. Consider that we have a scalar field φ and the electromagnetic field $F_{\mu\nu}$, as if they are non-interacting, given by the Lagrangian

$$\mathscr{L} = \frac{1}{4} F_{\mu\nu} F^{\mu\nu} + \eta^{\mu\nu} \varphi^*_{,\mu} \varphi_{,\nu} - U(\varphi) \tag{7.18}$$

The first term is just the electromagnetic Lagrangian, the second term is the kinetic term of the scalar field φ, and the third term is the potential energy of φ chosen to be a generalization of the quartic scalar potential seen in the previous chapter

$$U(\varphi) = -2\alpha\beta\varphi^*\varphi + \alpha^2(\varphi^*\varphi)^2$$

where α, β are constants. Notice that there is not an explicit interaction term involving the two fields, so they behave as if they were two independent fields.

As we have seen in the last section, the electromagnetic field is invariant under the gauge transformation (7.8) and (7.9), which can now be written in terms of the components of the four-vector potential as

$$A'_\mu = A_\mu + \theta(x)_{,\mu}$$

On the other hand, from the same arguments seen in the study of the quartic potential in the previous chapter, the scalar field component in the Lagrangian is invariant under the global $U(1)$ transformations, but not under the local $U(1)$ group given by the transformations

$$\varphi' = e^{i\theta'(x)}\varphi$$

Therefore, the Nielsen–Olesen Lagrangian also has two independent local gauge groups: the gauge group of the electromagnetic field with parameter $\theta(x)$ and the unitary gauge group $U(1)$ of the scalar field with parameter $\theta'(x)$. As we have seen, the latter gauge transformation is not a symmetry, unless we take infinitesimal transformations like in (6.12), and replacing the partial derivatives in the Lagrangian by the more general covariant derivative (6.13) $D_\mu = I\partial_\mu + i\theta'_{,\mu}$.

To solve the problem of handling two independent gauge transformations Nielsen and Olesen proposed that they are different manifestations of the same group, by assuming that

$$i\theta'_{,\mu} = gA_\mu, \quad g = \text{constant} \tag{7.19}$$

With such condition, the Lorentz gauge implies that $\partial^\mu\theta'_{,\mu} \equiv g\partial^\mu A_\mu = 0$. Therefore, using the Lorentz gauge, the two gauge symmetries become just one, namely $U(1)$, and the gauge covariant derivative becomes

$$D_\mu = \partial_\mu + i\theta'_{,\mu} = \partial_\mu + gA_\mu \tag{7.20}$$

Then the original Lagrangian can be rewritten with D_μ in place of the partial derivative ∂_μ:

$$\mathscr{L} = \frac{1}{4}F_{\mu\nu}F^{\mu\nu} + \eta^{\mu\nu}(D_\mu\varphi)^*(D_\nu\varphi) - U(\varphi) \tag{7.21}$$

With this covariant derivative the Lagrangian becomes constant under the local gauge $U(1)$.

Since D_μ depends on A_μ, the original Nielsen–Olesen Lagrangian acquired an interaction term that did not exist before. To see this term explicitly, let us expand the covariant derivatives

$$\mathscr{L} = \frac{1}{4}F_{\mu\nu}F^{\mu\nu} + \eta^{\mu\nu}\left(\varphi^*_{,\mu}\varphi_{,\nu} + gA_\mu\varphi^*\varphi_{,\nu} + gA_\nu\varphi^*_{,\mu}\varphi + g^2A_\mu A_\nu\varphi^*\varphi\right) - U(\phi)$$

where the interaction terms are those involving products of both fields and their derivatives. The emergence of the interaction has some interesting physical consequence.

7.1.3.1 The Meissner Effect

The Euler–Lagrange equations obtained from (7.21) with respect to A_μ are

$$F^{\mu\nu}{}_{,\mu} - g\,\eta^{\mu\nu}[\varphi^*_{,\nu}\varphi + \varphi_{,\nu}\varphi^* + 2gA_\nu\varphi^*\varphi] = 0 \tag{7.22}$$

and with respect to φ they are

$$g\eta^{\mu\nu}\left[A_\mu(D_\nu\varphi)^* - \frac{\partial U(\varphi)}{\partial\varphi} - g\eta^{\mu\nu}D_\nu(\varphi\varphi^*)\right] = 0 \tag{7.23}$$

(here we have written these equations using D_μ just for convenience. The Euler–Lagrange equations are usually written with the ordinary derivatives.) Since A_μ are the components of the electromagnetic potential, it must also satisfy the dynamical Maxwell's equations (7.17). Therefore, replacing the Maxwell tensor $F_{\mu\nu}$ in (7.22), we obtain a total of eight equations and only five unknowns A_μ and φ, so that the system is over-determined. This means that we cannot guarantee that the system remains consistent in its evolution.

The excess of equations can be lessened by reducing the number of dimensions from 4 to 3 = 2 + 1 (with coordinates x,y,t). There is no fundamental implication in this, as it means only that the sought solution is valid only in a three-dimensional subspace-time of space–time. In this case we obtain a compatible system with five equations. From (7.22) a solution of this system on empty space ($J^\mu = 0$), is given by

$$A_\mu = -\frac{(\varphi\varphi^*_{,\mu} - \varphi^*\varphi_{,\mu})}{2g\,\varphi\varphi^*} \tag{7.24}$$

Replacing this in (7.23) we obtain an equation involving only $\varphi(x, y, t)$.

In a practical application of this solution, consider that S is a flat surface limited by a circle c with radius r, in a region where there is no electrical current. Using the center of the circle as the center of a polar system of coordinates (r, θ, t) we may express the solution $\varphi = \sqrt{f(r)}e^{i\theta}$. Replacing this solution in (7.22) we obtain the electromagnetic potential A_μ in terms of $f(r)$ and θ.

We may choose the radius of the circle such that $\varphi^*\varphi = 1$. In this normalization, the magnetic field generated by this (2+1)-potential vector, as always, given by $B = \nabla \wedge A$, produces a *magnetic flux* across an arbitrary surface S in the (x, y) plane, limited by a closed curve c, given by

$$\Phi = \int\int < B, dS_n > = \int\int < \nabla \wedge A, dS_n > = \oint_c A_\mu dx^\mu$$

In the traditional electromagnetic field, this flux would be given by a current in c. Since here we have a vacuum solution ($J^\mu = 0$), then the flux should also be zero. However, using the above solution we find that

$$\Phi = \oint A_\mu dx^\mu = \frac{n}{g} \oint \theta_{,\mu} dx^\mu = \frac{n\pi}{g}$$

where n is the number of times in which the circle is run. *Contrary to the expectations it is not zero,* but it is discrete, depending on this integer n.

This result was confirmed by an experiment by Walther Meissner and Robert Ochsenfeld in 1933 and is known as the *Meissner effect* [96]. The circle c was drawn in a neutral metal plate (without any electrical current). A coil with n turns (*called the winding number*) with the same diameter as the circle was placed orthogonally to the plate. When a current flows in the coil, a magnetic flux should be produced on the disk drawn in the plate, but classically and at room temperature that flux is shielded by the plate itself. Nonetheless, at the critical temperature, they observed a flux distribution on the opposite side of the plate. The only possible interpretation of this somewhat strange result is that of a tunneling effect of a quantum *magnetic flux*. Within the assumptions made, the quantum effect on the flux appears under extremely low temperatures. When the temperature rises the quantum flux disappears.

When the plate is kept at room temperature and the magnetic field is produced by a cooled permanent magnet, then the flux causes a levitating effect on the magnet. The Meissner effect is thus responsible for the ongoing experiments on magnetic levitation and applications in public transportation.

The existence of a quantized flux only on one side of the plane may be also interpreted as the result of quantum *magnetic monopole* called the 'tHooft-Polyakov monopole [97, 98]. As in the example given by (6.1), the corresponding magnetic charge can be obtained by a symmetry breaking mechanism. More specifically, consider (7.21) where the parameters are chosen to be $\alpha^2 = \lambda/3! > 0$ and $2\alpha\beta = \mu^2$. Then the minimal energy condition $\partial U/\partial\varphi = 0$ gives

$$\varphi^* \left(\mu^2 + \frac{\lambda}{3!}(\varphi^*\varphi) \right) = 0$$

Therefore, if $\mu^2 > 0$, the only solution $\varphi = 0$. On the other hand, if $\mu^2 = -m^2 < 0$, then we have an infinite number of non-trivial vacuum states given by

$$\varphi_0 = \pm\sqrt{\frac{6m^2}{\lambda}} e^{i\theta}$$

When any of these infinite vacua states is chosen for a value of θ, the gauge symmetry of the field φ is broken and the Lagrangian acquires a mass term m^2 (proportional to $\varphi^*\varphi$), which can be interpreted as the *magnetic mass of the monopole*.

7.2 Spinor Fields

The best way to define *spinor fields* is through a particular tensor structure called a Clifford algebra defined on space–time.

Definition 7.1 (Clifford Algebras) The Clifford algebra $\mathbb{C}_n$ generated by an n-dimensional vector space V is the quotient of the tensor algebra $V \otimes V$ by the bilateral ideal I, defined by a bilinear form B in V and denoted by [99]

$$\mathbb{C}_n = (V \otimes V)/I$$

A bilinear form is a map $B : V \times V \to I\!R$, which is linear in both arguments $B(\mathbf{v}, \mathbf{w}) \in I\!R$. The above expression defines a subspace of the tensor algebra $V \otimes V$ given by the condition

$$\mathbf{v} \otimes \mathbf{w} + \mathbf{w} \otimes \mathbf{v} = B(\mathbf{v}, \mathbf{w})$$

This specifies that the rank-2 tensors in $\mathbb{C}_n$ are symmetric tensors $(V \otimes V)$ and that they are proportional to $B(\mathbf{v}, \mathbf{w})$. In terms of a basis $\{e_\alpha\}$ of V the bilateral ideal corresponds to imposing to the tensor algebra the condition

$$\mathbf{e}_\alpha \otimes \mathbf{e}_\beta + \mathbf{e}_\beta \otimes \mathbf{e}_\alpha = B(\mathbf{e}_\alpha, \mathbf{e}_\beta)$$

In general the tensor product notation in $\mathbb{C}_n$ is simplified to $e_\alpha \otimes e_\beta + e_\beta \otimes e_\alpha = e_\alpha e_\beta + e_\beta e_\alpha$. Denoting the coefficients of the bilinear form by $B(\mathbf{e}_\alpha, \mathbf{e}_\beta) = 2g_{\alpha\beta}$, we may write the Clifford algebra as

$$e_\alpha e_\beta + e_\beta e_\alpha = 2g_{\alpha\beta} e_0 \tag{7.25}$$

where e_0 denotes the identity element of the algebra:

$$e_\alpha e_0 = e_0 e_\alpha$$

The dimension of $\mathbb{C}_n$ is given by the maximum number 2^n of linearly independent elements of the algebra obtained with the independent products of the generators. Therefore, a generic element of $\mathbb{C}_n$ is given by the linear combination of the generators and their independent products:

$$X = X^0 e_0 + X^\alpha e_\alpha + X^{\alpha\beta} e_\alpha e_\beta + \cdots + X^{\alpha\beta\ldots\gamma} e_\alpha e_\beta, \cdots, e_\gamma$$

Example 7.1 (Complex Algebra) The complex algebra is the simplest Clifford algebra $\mathbb{C}_1 = \mathbb{C}$, with just one generator $e_1 = i$ plus the identity element $e_0 = 1$. The dimension of the algebra is $2^1 = 2$, and its generic elements are like

$$X = X_0 e_0 + X_1 e_1 = X_0 1 + X_1 i$$

It is usual to consider the real $I\!R$ as a Clifford algebra with just the identity element, denoted by $\mathbb{C}_0 = I\!R$.

After the complex algebra, the better known Clifford algebra is the quaternion algebra (or hypercomplex algebra) defined by William Hamilton in 1843 [100].

Example 7.2 (Quaternions) The quaternion algebra is the Clifford algebra $\mathbb{C}_2$ generated by a two-dimensional vector space.

Denoting by $\{e_\alpha\}$ an orthonormal basis of the three-dimensional space, with metric coefficients $\delta_{\alpha\beta}$, the quaternion algebra is given by the multiplication table

$$\begin{aligned} e_1 e_2 + e_2 e_1 &= 0 \\ e_1 e_0 = e_1, \quad e_2 e_0 &= e_2 \\ e_1 e_1 = e_2 e_2 &= -e_0 \end{aligned}$$

Denoting $e_3 = e_1 e_2$ and $X_{12} = X_3$, the quaternion can be written as

$$X = X^0 e_o + X^1 e_1 + X^2 e_2 + X^3 e_3$$

and the multiplication table can be simplified to

$$e_\alpha e_\beta + e_\beta e_\alpha = -2\eta_{\alpha\beta} e_0, \quad e_0 e_\alpha = e_\alpha e_0 = e_\alpha, \quad \alpha, \beta, \dots 1..3 \tag{7.26}$$

The conjugate of a quaternion is defined by

$$\bar{e}_\alpha = -e_\alpha, \quad \bar{e}_0 = e_0$$

and the norm of a quaternion is

$$||X||^2 = X\bar{X} = X_0^2 + X_1^2 + X_2^2 + X_3^2$$

It should be mentioned that the complex and the quaternion algebras are the only associative normed division algebras, that is, such that $||AB|| = ||A||\,||B||$ and $(AB)C = A(BC)$ (by extension the set of real numbers is considered as a Clifford algebra generated by the identity only). The division algebra property is relevant to the construction of the standard mathematical analysis based on the properties of limits and derivatives, allowing us to write

$$\lim_{\Delta x \to 0} ||\frac{\Delta F(x)}{\Delta x}|| = \lim_{\Delta x \to 0} \frac{||\Delta F(x)||}{||\Delta x||}$$

The set of real numbers $I\!R$ is a division algebra because we have the same division property, where in fact the concept appeared in the first place. There is a fourth division algebra called the octonion algebra with seven generators, although it is not associative. We shall return to it at the end in connection with the $SU(3)$ gauge theory.

Definition 7.2 (Spinors) Spinors are vectors of a representation of the group of automorphisms of a Clifford algebra defined on space–time, satisfying a given variational principle:

Given an algebra $\mathscr{A}$, an n-dimensional matrix representation of it is a homomorphism

$$\mathscr{R} : \mathscr{A} \rightarrow M_{n\times n}$$

where $M_{n\times n}$ denotes the $n \times n$ matrix algebra. Denoting by $\mathscr{R}(X)$ and $\mathscr{R}(Y)$ the matrix representing $X, Y \in \mathscr{A}$, the homomorphism condition says that the product of the algebra goes into the product of matrices $\mathscr{R}(XY) = \mathscr{R}(X)\mathscr{R}(Y)$.

Any matrix representation of an algebra can be seen as linear operators on some vector space $\mathscr{S}$, whose vectors are represented by a column

$$\varphi = \begin{pmatrix} \varphi_1 \\ \varphi_2 \\ \vdots \\ \varphi_N \end{pmatrix} \tag{7.27}$$

In particular, we may construct spinor representations of Clifford algebras defined on a space–time. The basic example is the representations of the quaternion algebra given by the Pauli matrices associated with the spin properties of particles in quantum mechanics [101]. The Pauli matrices can be written in a variety of ways, corresponding to equivalent representations. Here we use the following:

$$\sigma_0 = \begin{pmatrix} 1 & 0 \\ 0 & 1 \end{pmatrix}, \quad \sigma_1 = \begin{pmatrix} 0 & 1 \\ 1 & 0 \end{pmatrix}, \quad \sigma_2 = \begin{pmatrix} 0 & -i \\ i & 0. \end{pmatrix}, \quad \sigma_3 = \begin{pmatrix} 1 & 0 \\ 0 & -1 \end{pmatrix} \tag{7.28}$$

such that they satisfy the multiplication table

$$\begin{cases} \sigma_i\sigma_j + \sigma_j\sigma_i = -2\delta_{ij}\sigma_0 \\ \sigma_0\sigma_i = \sigma_i\sigma_0 \\ \sigma_0\sigma_0 = \sigma_0 \end{cases}$$

which is the same multiplication table of the quaternion algebra.

The above representation can be used as the basis to construct a matrix representation of any Clifford by tensor products of matrices, known as the Brauer–Weyl representation or simply as the Weyl representation [102]:

$$\begin{cases} P_\alpha = \sigma_2 \otimes \sigma_2 \otimes \cdots \otimes \sigma_2 \otimes \sigma_1 \otimes \sigma_0 \otimes \cdots \otimes \sigma_0 \otimes \sigma_0 \\ P_{n+1} = \sigma_2 \otimes \sigma_2 \otimes \cdots\cdots\cdots\cdots\cdots\cdots\cdots \otimes \sigma_2 \otimes \sigma_2 \\ Q_\alpha = \sigma_2 \otimes \sigma_2 \otimes \cdots \otimes \sigma_2 \otimes \sigma_3 \otimes \sigma_0 \otimes \cdots \otimes \sigma_0 \otimes \sigma_0 \end{cases}$$

where the matrices σ_1 and σ_3 occupy the α position. The tensor product $\otimes$ is taken to be from left to right (that is, each entry of the left matrix is multiplied by the whole right matrix).[1]

The column vectors (7.27) of the representation space $\mathscr{S}$ of a matrix representation of a Clifford algebra are called *spinors*. From the above Brauer–Weyl representations we may conclude that the spinors of a representation of $\mathbb{C}_n$ with n generators $\{\mathbf{e}_\alpha\}$ have $N = 2^{[n]/2}$ independent components, where $[n] = n$ for even n and $[n] = n - 1$ for odd n.

An important result shows that $\mathbb{C}_{2n+1} \approx \mathbb{C}_{2n}/\mathbb{C}$, where the right-hand side denotes the Clifford algebra on the complex field (with complex coefficients). Thus, the Dirac matrices in five dimensions are essentially the same as Dirac matrices in four dimensions.

An interesting case occurs in eight dimensions, where the spinors have $2^{8/2} = 16$ components, but they split in two equivalent halves with eight components each [103]. If in addition these spinors are real, then each half spinor space is isomorphic to the generator space of the Clifford algebra.

Example 7.3 (Pauli Spinors) The Pauli matrices (7.28) define two-component spinor representation of the quaternion algebra. Indeed, the quaternion algebra is the Clifford algebra $\mathbb{C}_2$ with two generators in the case of a two-dimensional (complex) spinor representation $\mathscr{S}_2$. Thus, we obtain a two-dimensional spinor field in $\mathscr{M}$, defined by

$$\Psi : \mathscr{M} \rightarrow \mathscr{S}_2$$

which gives a two-component spinor at each point of the space–time

$$\Psi(p) = \begin{pmatrix} \Psi_1 \\ \Psi_2 \end{pmatrix}_p$$

[1] Tensor products in general are non-commutative. Here, Brauer and Weyl prescribed a specific way to do it. It is possible to reverse the order, obtaining a different representation. Other spinor representations, such as the Majorana and Majorana–Weyl, different from the one above are also used in field theory.

Example 7.4 (Dirac Spinors) The Dirac spinors are vectors of the four-dimensional spinor representation of the Clifford algebra $\mathbb{C}_4$, generated by a four-dimensional space.

Taking the generating space to be Minkowski's space–time $\{e_\mu\}$, together with the identity element e_0, we obtain an algebra with 16 components, whose general element is written as

$$X = X^0 e0 + X^\alpha e_\alpha + X^{\alpha\beta} e_\alpha e_\beta + X^{\alpha\beta\gamma} e_\alpha e_\beta e_\gamma + \cdots + X^{1234} e_1 e_2 e_3 e_4$$

The Brauer–Weyl matrix representation (simply known as the Weyl representation) of this algebra gives the $2^{[4]/2} \times 2^{[4]/2}$ matrices which are the Dirac matrices

$$\gamma_1 = \begin{pmatrix} 0 & 0 & 0 & -i \\ 0 & 0 & i & 0 \\ 0 & i & 0 & 0 \\ -i & 0 & 0 & 0 \end{pmatrix}, \quad \gamma_2 = \begin{pmatrix} 0 & 0 & 0 & i \\ 0 & 0 & i & 0 \\ 0 & i & 0 & 0 \\ i & 0 & 0 & 0 \end{pmatrix}, \quad \gamma_3 = \begin{pmatrix} 0 & 0 & -i & 0 \\ 0 & 0 & 0 & i \\ i & 0 & 0 & 0 \\ 0 & -i & 0 & 0 \end{pmatrix}, \quad \gamma_4 = \begin{pmatrix} 0 & 0 & i & 0 \\ 0 & 0 & 0 & i \\ -i & 0 & 0 & 0 \\ 0 & -i & 0 & 0 \end{pmatrix}$$

These matrices act as linear operators on a four-dimensional complex space V_4, which is the Dirac spinor space in the Minkowski space–time

$$\psi = \begin{pmatrix} \psi_1 \\ \psi_2 \\ \psi_3 \\ \psi_4 \end{pmatrix}$$

satisfying Dirac's equation for a relativistic charged particle with spin $1/2$ and mass m

$$(\gamma^\mu \partial_\mu - m)\psi = 0 \tag{7.29}$$

This equation can be derived from the Dirac Lagrangian [104]

$$\mathscr{L} = \bar{\psi}(\gamma^\mu \partial_\mu + m)\psi \tag{7.30}$$

where we have denoted $\bar{\psi} = \psi^\dagger \gamma^5$, and where $\psi^\dagger = (\psi^T)^*$ and $\gamma^5 = \gamma_1\gamma_2\gamma_3\gamma_4$.

7.2.1 Spinor Transformations

Since spinor fields are derived from representations of Clifford algebras, the (internal) automorphisms of these algebra correspond to a spinor transformation, that is, given a map $\tau : \mathbb{C}_n \to \mathbb{C}_n$ defined by $e'_\mu = \tau e_\mu \tau^{-1}$, such that it maintains invariant the multiplication table

$$\tau e_\mu e_\nu \tau^{-1} + \tau e_\nu e_\mu \tau^{-1} = g_{\mu\nu} e_0$$

It follows that these automorphisms necessarily correspond to isometries of the metric in the generating space. In the case of a matrix representation, the operators τ correspond to a matrix acting on the spinor space as

$$\Psi' = \mathscr{R}(\tau)\Psi$$

In quantum mechanics, spinors represent quantum states and therefore these matrix transformations $\mathscr{R}(\tau)$ correspond to unitary matrix operators denoted by u, such that $uu^\dagger = 1$ and acting on the spinors as $\Psi' = u\Psi$.

Example 7.5 (Transformations of Dirac Spinors) Let us detail the transformation of the Lagrangian of the Dirac spinor field under a unitary gauge transformation $\Psi' = u\Psi$, of the local group $U(1)$: $u = e^{i\theta(x)} e_0$. The derivative of the transformed spinor gives

$$\Psi'_{,\mu} = e^{i\theta(x)} \Psi_{,\mu} + i\theta_{,\mu} e^{i\theta(x)} \Psi$$

and similarly for $\bar{\Psi}'$. Replacing these transformations in the Dirac Lagrangian (7.30) we find that

$$\begin{aligned} \mathscr{L}(\Psi') &= e^{-i\theta(x)} \bar{\Psi}[\gamma^\mu (e^{i\theta} \Psi_{,\mu} + i\theta_{,\mu} e^{i\theta} \Psi)] - m\bar{\Psi}\Psi \\ &= \bar{\Psi}(\gamma^\mu \partial_\mu - m)\Psi + i\theta_{,\mu} \bar{\Psi}\Psi \end{aligned}$$

We see clearly that the Lagrangian is not invariant due to the presence of the derivative of the parameter θ. However, as it happened in the case of the complex scalar field, taking an infinitesimal transformation and defining the covariant derivative

$$\gamma^\mu D_\mu = \gamma^\mu \partial_\mu + i\theta_{,\mu}$$

Then the Lagrangian becomes invariant:

$$\mathscr{L}(\psi') = \bar{\psi}(i\gamma^\mu D_\mu - m)\Psi = \mathscr{L}(\psi)$$

Example 7.6 (Isospin) Returning to the quaternion algebra $\mathbb{C}_2$ satisfying the multiplication table (7.26), we have an algebra that is invariant under the group of rotations $SO(3)$ (that can be seen as a subgroup of the Galilei group in $\mathscr{G}_4$). When this algebra is represented by the Pauli matrices (7.28), the corresponding quantum states describe the orbital spin states.

On the other hand, we also have an internal action of the same algebra, but which has nothing to do with the rotations in space–time. The matrix representations of this global automorphism produce two-component spinors called isospin, which transforms as

$$\begin{pmatrix} \Psi_1' \\ \Psi_2' \end{pmatrix} = e^{i\Theta} \begin{pmatrix} \Psi_1 \\ \Psi_2 \end{pmatrix}$$

where now θ is the parameter of the global gauge symmetry. The matrix representation of the automorphism is the same as Pauli matrices, but to avoid confusion we use a different notation

$$\tau_0 = \begin{pmatrix} 1 & 0 \\ 0 & 1 \end{pmatrix}, \quad \tau_1 = \begin{pmatrix} 0 & 1 \\ 1 & 0 \end{pmatrix}, \quad \tau_2 = \begin{pmatrix} 0 & -i \\ i & 0 \end{pmatrix}, \quad \tau_3 = \begin{pmatrix} 1 & 0 \\ 0 & -1 \end{pmatrix}$$

Chapter 8
Noether's Theorem

A major contribution to the development of the physics of fundamental interactions, in fact for the whole physics, was the theorem developed by Emmy Noether in 1918 [10, 105, 106], showing how to construct the observables of a theory, given its Lagrangian and Lie symmetry groups. Following Lopes, for pedagogical purposes we divide the theorem in three parts, one for each kind of symmetry considered: transformation of coordinates in space–time, global gauge transformations, and local gauge transformations [107]. We also include separately the specific case of general relativity with its peculiar diffeomorphism invariance. In all cases we will denote by $\Psi(x)$ some field (scalar, vector, tensor, spinor) defined in space–time $\mathscr{M}$, satisfying the Euler–Lagrange field equations derived from a Lagrangian $\mathscr{L}$, generally depending on $\Psi(x)$ and its first derivatives

$$\mathscr{L} = \mathscr{L}(\Psi, \Psi_{,\lambda})$$

8.1 Noether's Theorem for Coordinate Symmetry

The coordinate symmetry of a field $\Psi(x)$ defined in any space–time specifies the transformation of coordinates which leaves the field equations in the same form. This is in fact the very basic origin of symmetry which becomes important after the discovery of the Lorentz/Poincaré transformations. Here and in all subsequent cases we will be using Lie symmetry groups and the corresponding Lie algebra. Therefore, it is sufficient to consider an infinitesimal coordinate transformation like

$$x'^{\mu} = x^{\mu} + \xi^{\mu} \tag{8.1}$$

where ξ^{μ} is the descriptor introduced in (8.1).

Coordinate transformations can be interpreted in two different ways: as a passive change of coordinates of the same point, for example, from Cartesian coordinates to spherical coordinates and as an active coordinate transformation, meaning transformation between coordinates of different points, say resulting from a motion. From the point of view view of tangent bundles, the first case represents a map in each

M.D. Maia, *Geometry of the Fundamental Interactions*,
DOI 10.1007/978-1-4419-8273-5_8,

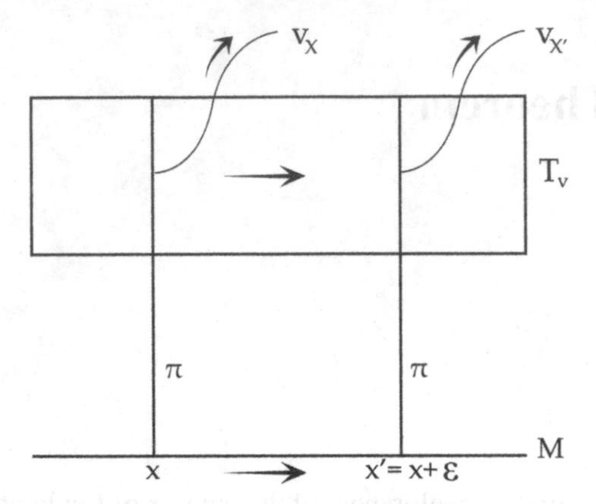

Fig. 8.1 Map between fibers from a coordinate transformation

fiber over the same point of $\mathscr{M}$. In the latter case, we have a map between two different points of $\mathscr{M}$ as shown in Fig. 8.1.

Depending on the nature of the field Ψ, the coordinates transformation may produce two kinds of variations.

(a) The *functional variation* which takes into consideration only the functional dependence of the field

$$\delta_F \Psi = \Psi(x') - \Psi(x)$$

In particular this applies to all scalar fields, and also for each individual component of any vector, tensor, or spinor field. Thus, the infinitesimal *functional variation* δ_F of any field under the infinitesimal coordinate transformation can be obtained from the Taylor expansion of $\Psi(x+\xi)$, keeping only the first power of θ:

$$\delta_F \Psi = \sum \delta\theta^a a_a^\mu(x) \frac{\partial \Psi}{\partial x^\mu} \tag{8.2}$$

The designation of functional variation is used also for the variation of a function of field variables, such as the Lagrangian. In this case, the functional analysis may require a more complicated study on the topology of open sets defined in the space of the field in question. However, in most applications including that of gauge fields on a manifold, the topological basis comes from the coordinate space. Therefore the derivative resumes the application of simple chain rule for derivatives. This, of course, depends on the field structure, as for example if it is commutative or not.

(b) The *algebraic variation* occurs when the field has an explicit algebraic nature, such as vector, tensor, or spinor. In this case, we need also to evaluate the change of the components of the field Ψ which is independent of the functional variation, but it depends on the nature of each type of field. We will denote this variation generically by

$$\delta_A \Psi = \Psi'(x) - \psi(x)$$

Since this algebraic variation depends on each type of field, we may start by examples.

The algebraic variation of a scalar field is by its own definition equal to zero.

On the other hand, the algebraic variation of a vector field A given by its contravariant components is given by the contravariant tensor transformation. For an infinitesimal transformation like in (8.1)

$$A'^{\mu} = \frac{\partial x'^{\mu}}{\partial x^{\rho}} A^{\rho} = (\delta^{\mu}_{\rho} + \xi^{\mu}_{,\rho}) A^{\rho} = \sum \delta\theta^{a} a^{\mu}_{a,\rho}(x) A^{\rho} = x'^{\mu}$$

Therefore,

$$\delta_A A^{\mu} = \sum \delta\theta^{a} a^{\mu}_{a,\rho}(x) A^{\rho} \tag{8.3}$$

Similarly, for a rank-2 contravariant tensor field we have for an infinitesimal coordinate transformation

$$\delta_A T^{\mu\nu} = \frac{\partial x'^{\mu}}{\partial x^{\rho}} \frac{\partial x'^{\nu}}{\partial x^{\sigma}} T^{\rho\sigma} = \sum \delta\theta^{a} (a^{\nu}_{a,\rho}(x) T^{\rho\mu} + a^{\nu}_{a,\sigma}(x) T^{\mu\sigma}) \tag{8.4}$$

As a last example consider the transformation of spinor fields. Since by definition the spinors are objects of the representation space of the Clifford algebra, they transform as

$$\Psi' = \tau(\theta)\Psi$$

where τ is a transformation of the automorphisms of the algebra, corresponding to the coordinate transformation. In particular for an infinitesimal coordinate transformation with infinitesimal parameters $\delta\theta^{a}$, the transformation of Ψ is a deviation from the identity I

$$\tau(\delta\theta)\Psi \approx \left(I + \sum a_a(x)\delta\theta^{a} \right) \Psi$$

so that the infinitesimal algebraic transformations of a spinor field are

$$\delta_A \Psi = \tau(\delta\theta)\Psi - \Psi = \sum a_a(x)\delta\theta^{a}\Psi \tag{8.5}$$

From examples (8.3), (8.4), and (8.5), we conclude that in general the algebraic variation of any field under coordinate transformations is a function of the coordinates, of the field and it is proportional to the parameters. This can be summarized as

$$\delta_A \Psi = \sum_a \mathcal{G}_a(x, \Psi)\delta\theta^{a} \tag{8.6}$$

where $\mathscr{G}_a$ denotes N functions of the coordinates and of the field, which are specified within a given transformation.

The *total variation* of the field is the result of the combined functional and algebraic changes, defined by

$$\delta_{\mathrm{T}}\Psi = \Psi'(x') - \Psi(x) \tag{8.7}$$

The total variation of a field corresponding to an infinitesimal coordinate transformation is given by

$$\delta_{\mathrm{T}}\Psi = \sum \mathscr{F}_a(x, \Psi, \Psi_{,\mu})\delta\theta^a \tag{8.8}$$

where $\mathscr{F}_a(x, \Psi, \Psi_{,\mu}), \quad a = 1 \ldots N$, are N functionals defined in the space of the field Ψ. In fact, starting from (8.7) we obtain

$$\delta_{\mathrm{T}}\Psi = \Psi(x') - \Psi(x) = \Psi'(x') - \Psi(x') + \Psi(x') - \Psi(x) = \delta_{\mathrm{A}}(x') - \delta_{\mathrm{F}}(x)$$

where we notice that the algebraic variation is the same in all points and that it is independent of the functional variation, that is, $\delta_{\mathrm{A}}\Psi(x') = \delta_{\mathrm{A}}\Psi(x)$. Consequently, using (8.2) and (8.6), the total variation is

$$\delta_{\mathrm{T}}\Psi(x) = \sum_a \left(\mathscr{G}_a(x, \Psi) + a_a^\mu(x)\Psi_{,\mu}\right)\delta\theta^a \tag{8.9}$$

The parenthesis is composed of functions of x, Ψ, and first derivatives of Ψ. Denoting $\mathscr{F}_a(x, \Psi, \Psi_{,\mu}) = (\mathscr{G}_a(x, \Psi)_a + a_a^\mu(x)\Psi_{,\mu})$, we obtain the expression (8.8) for the total variation. Notice that $\mathscr{F}_a$ has the same algebraic structure as if Ψ.

Consider the examples:

For a vector field A_μ, we have functional and algebraic variations so that the total variation is given by

$$\delta_{\mathrm{T}}A^\mu = A'^\mu(x') - A^\mu(x) = \sum \left(\frac{\partial a_a^\mu}{\partial x^\rho}A^\rho + \frac{\partial A^\mu}{\partial x^\rho}a_a^\rho\right)\delta\theta^a$$

Comparing with (8.8) we obtain the functions

$$\mathscr{F}_a^\mu(A^\mu, A_\rho^\mu, x^\mu) = \sum \left(\frac{\partial a_a^\mu}{\partial x^\rho}A^\rho + \frac{\partial A^\mu}{\partial x^\rho}a_a^\rho\right)$$

For a rank-2 tensor field the total variation from an infinitesimal coordinate transformation $x'^\mu = x^\mu + \xi^\mu$ is

$$\delta_{\mathrm{T}}T^{\mu\nu} = T'^{\mu\nu}(x') - T^{\mu\nu}(x) = T^{\mu\nu}{}_{,\tau}\xi^\tau + T^{\mu\sigma}\xi^\nu_\sigma + T^{\nu\sigma}\xi^\mu_{,\sigma}$$

Therefore we have

$$\mathscr{F}_a^{\mu\nu}(T, T_{,\lambda}, x) = \sum \left(\frac{\partial T^{\mu\nu}}{\partial x^{\tau}} a_a^{\tau} + T^{\mu\sigma} \frac{\partial a_a^{\nu}}{\partial x^{\sigma}} + T^{\nu\sigma} \frac{\partial a_a^{\mu}}{\partial x^{\sigma}} \right)$$

The procedure is similar for other tensors.

Theorem 8.1 (Noether's Theorem for Coordinate Symmetries) *Given the Lagrangian for a field Ψ defined in a closed region Ω of space–time with boundary $\partial\Omega$, such that it is invariant under an N-parameter infinitesimal coordinate transformation, then there are N quantities*

$$\mathscr{N}_a^{\lambda} = \frac{\partial \mathscr{L}}{\partial \Psi_{,\lambda}} (\mathscr{F}_a - \Psi_{,\mu} a_a^{\mu}) + \mathscr{L} a_a^{\lambda} \tag{8.10}$$

which are conserved in the sense that

$$\mathscr{N}^{\lambda}{}_{a,\lambda} = 0$$

Consider the action for the field Ψ for which we use the short notation

$$A(\Psi, \Omega) = \int_{\Omega} \mathscr{L}(\Psi(x)) dv$$

Under an infinitesimal coordinate transformation the action is scalar depending on the limits of integration and hence it has functional variation only:

$$\delta_{\mathrm{T}} A = A(\Psi', \Omega') - A(\Psi, \Omega)$$

where Ω' is the same region Ω except that it is described by the new coordinates x'^{μ}. After summing and subtracting $A(\Psi, \Omega')$, we obtain

$$\delta_{\mathrm{T}} A = A(\Psi', \Omega') - A(\Psi, \Omega') + A(\Psi, \Omega') - A(\Psi, \Omega)$$

or equivalently

$$\begin{aligned} \delta_{\mathrm{T}} A &= \int_{\Omega'} \mathscr{L}(\Psi'(x')) dv' - \int_{\Omega'} \mathscr{L}(\Psi(x') dv' + \int_{\Omega'} \mathscr{L}(\Psi(x')) dv' - \int_{\Omega} \mathscr{L}(\Psi(x)) dv \\ &= \int_{\Omega'} \left[\mathscr{L}(\Psi'(x')) - \mathscr{L}(\Psi(x')) \right] dv' + \int_{\Omega'} \mathscr{L}(\Psi(x')) dv' - \int_{\Omega} \mathscr{L}(\Psi(x)) dv \end{aligned} \tag{8.11}$$

The term in brackets represents the algebraic variation of $\mathscr{L}$ which is a scalar functional, depending on an algebraic field Ψ, so that there is an algebraic variation

$$\delta_{\mathrm{A}} \mathscr{L} = [\mathscr{L}(\Psi'(x')) - \mathscr{L}(\Psi(x'))]$$

On the other hand from elementary calculus we know that

$$\int_{\Omega'} f(x')dv' = \int_{\Omega} f(x) \det \mathscr{J} \, dv$$

where $\mathscr{J}$ is the Jacobian matrix of the infinitesimal coordinate transformation $x'^{\mu} = x^{\mu} + \xi^{\mu}$

$$\mathscr{J}(x, x') = \left(\frac{\partial x'^{\mu}}{\partial x^{\nu}} \right) = \left(\delta^{\mu}_{\nu} + \frac{\partial \xi^{\mu}}{\partial x^{\nu}} \right)$$

and

$$\det \mathscr{J} = 1 + \delta^{\mu}_{\nu} \frac{\partial \delta x^{\mu}}{\partial x^{\nu}} = 1 + \sum \frac{\partial \xi^{\mu}}{\partial x^{\mu}}$$

Therefore $dv' = \left(1 + \sum \partial \xi^{\mu} / \partial x^{\mu}\right) dv$ and the total variation (8.11) is

$$\delta_{\mathrm{T}} A = \int_{\Omega} \delta_{\mathrm{A}} \mathscr{L}(\Psi(x)) \left(1 + \sum \frac{\partial \xi^{\mu}}{\partial x^{\mu}}\right) dv + \int_{\Omega} \mathscr{L}(\Psi(x)) \left(1 + \frac{\xi^{\mu}}{\partial x^{\mu}}\right) - \int_{\Omega} \mathscr{L}(\Psi(x)) dv$$

$$= \int_{\Omega} \left[\delta_{\mathrm{A}} \mathscr{L}(\Psi(x)) + \delta_{\mathrm{A}} \mathscr{L}(\Psi(x)) \frac{\partial \xi^{\mu}}{\partial x^{\mu}} + \mathscr{L}(\Psi(x)) \frac{\partial \xi^{\mu}}{\partial x^{\mu}} \right] dv$$

As we commented before, the algebraic variation of the Lagrangian functional is not zero and from (8.6) it is proportional to $\delta\theta$. Therefore, the second term in the above integral is quadratic in $\delta\theta^{a}$ (the descriptor ξ is proportional to $\delta\theta$) and consequently it can be neglected. Consequently

$$\delta_{\mathrm{T}} A = \int_{\Omega} \left[\delta_{\mathrm{A}} \mathscr{L}(\Psi(x)) + \mathscr{L}(\Psi(x)) \frac{\partial \delta \xi^{\mu}}{\partial x^{\mu}} \right] dv$$

Now, $\delta_{\mathrm{A}} \mathscr{L} = \dfrac{\partial \mathscr{L}}{\partial \Psi} \delta_{\mathrm{A}} \Psi + \dfrac{\partial \mathscr{L}}{\partial \Psi_{,\lambda}} \delta_{\mathrm{A}} (\Psi_{,\lambda})$ so that

$$\delta_{\mathrm{T}} A = \int_{\Omega} \left[\frac{\partial \mathscr{L}}{\partial \Psi} \delta_{\mathrm{A}} \Psi + \frac{\partial \mathscr{L}}{\partial \Psi_{,\lambda}} \delta_{\mathrm{A}} (\Psi_{,\lambda}) + \mathscr{L}(\Psi) \frac{\partial \xi^{\mu}}{\partial x^{\mu}} \right] dv$$

Using the Euler–Lagrange equations obtained from the variational principle (remembering that in the derivation of these equations the variation of Ψ vanishes at the boundary $\partial \Omega$)

$$\frac{\partial \mathscr{L}}{\partial \Psi} = \frac{\partial}{\partial x^{\lambda}} \left(\frac{\partial \mathscr{L}}{\partial \Psi_{,\lambda}} \right)$$

Therefore,

$$\delta_{\mathrm{T}} A=\int_{\Omega}\left[\frac{\partial}{\partial x^{\lambda}}\left(\frac{\partial \mathscr{L}}{\partial \Psi_{,\lambda}}\right) \delta_{\mathrm{A}} \Psi+\frac{\partial \mathscr{L}}{\partial \Psi_{,\lambda}} \delta_{\mathrm{A}} \Psi_{,\lambda}+\frac{\partial \mathscr{L}}{\partial x^{\lambda}} \delta x^{\lambda}+\mathscr{L}(\Psi) \frac{\partial}{\partial x^{\lambda}}(\xi^{\lambda})\right] d v$$

or

$$\delta_{\mathrm{T}} A=\int_{\Omega} \frac{\partial}{\partial x^{\lambda}}\left[\frac{\partial \mathscr{L}}{\partial \Psi_{,\lambda}} \delta_{\mathrm{A}} \Psi+\mathscr{L} \xi^{\lambda}\right] d v \tag{8.12}$$

where we have used the fact that the variation δ is independent of the partial derivative $\delta_{\mathrm{A}} \Psi_{,\lambda}=(\partial / \partial x^{\lambda}) \delta_{\mathrm{A}} \Psi$. Using the definition $\xi^{\lambda}=\sum a_{a}^{\lambda} \delta \theta^{a}$ and from the comparison between (8.8) and (8.9) we may write $\delta_{\mathrm{A}} \Psi=\left(\mathscr{F}_{a}-\Psi_{,\lambda} a_{a}^{\lambda}\right) \delta \theta^{a}$. Therefore,

$$\delta_{\mathrm{T}} A=\int_{\Omega} \frac{\partial}{\partial x^{\lambda}}\left[\frac{\partial \mathscr{L}}{\partial \Psi_{,\lambda}}(\mathscr{F}_{a}-\Psi^{,\lambda} a_{a}^{\lambda})+\mathscr{L} a_{a}^{\lambda}\right] \delta \theta^{a} d v$$

Denoting

$$\mathscr{N}_{a}^{\lambda}=\frac{\partial \mathscr{L}}{\partial \Psi_{,\lambda}}(\mathscr{F}_{a}-\Psi_{,\mu} a_{a}^{\mu})+\mathscr{L} a_{a}^{\lambda} \tag{8.13}$$

and considering that the variational principle is maintained along the total variation $\delta_{\mathrm{T}} A=0$, we obtain the equation

$$\int_{\Omega} \delta \theta^{a} \frac{\partial}{\partial x^{\lambda}} \mathscr{N}_{a}^{\lambda} d v=0$$

However, $\delta \theta^{a}$ are linearly independent parameters (which are also independent of x^{μ}), so that

$$\int_{\Omega} \frac{\partial}{\partial x^{\lambda}} \mathscr{N}_{a}^{\lambda} d v=0$$

Therefore, if Ω is a closed region, applying the divergence theorem to $\mathscr{N}_{a}$, we obtain the null divergence of $\mathscr{N}_{a}$

$$\frac{\partial}{\partial x^{\lambda}} \mathscr{N}_{a}^{\lambda}=\mathscr{N}_{a, \lambda}^{\lambda}=0$$

which proves the theorem.

To understand this result consider in particular that Ω is limited by two space-like hypersurfaces, S_1 and S_2, close to each other and a time-like hypersurface S_3 (Fig. 8.2).

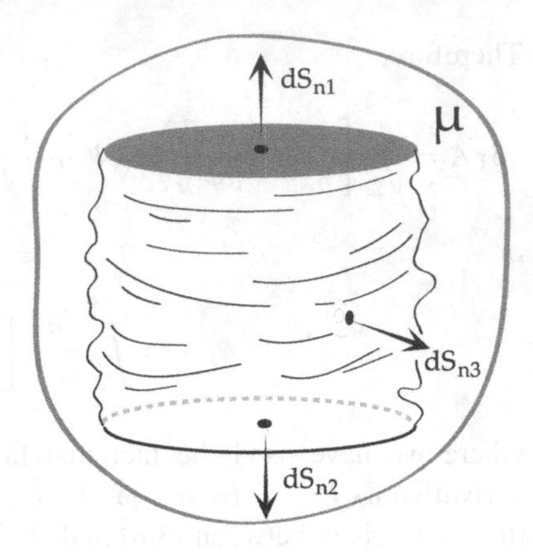

Fig. 8.2 Boundaries of a closed region in $\mathcal{M}$

$$\int_{\Omega} \mathcal{N}^{\lambda}_{a,\lambda} dv = \int_{S} < \mathcal{N}_a, dS_n >= \int_{S_1} \mathcal{N}^1_a dS_1 + \int_{S_2} \mathcal{N}^2_a dS_2 + \int_{S_3} \mathcal{N}^3_a dS_3 = 0$$

Under the conditions in which the variational principle was defined, the integral over S_3 vanishes. After reorienting S_1, it follows that the quantity

$$\int_{S_1} \mathcal{N}^1_a dS_1 = \int_{S_2} \mathcal{N}^2_a dS_2$$

is independent of the surface S. In other words, it is the same along the evolution of the field.

The object $\mathcal{N}^{\lambda}_a$, called the *Noether quantity*, belongs to the same algebra in which the field Ψ is defined. It is defined up to the addition of a term proportional to a tensor with zero divergence. In fact if $T^{\lambda\nu}_a$ is such that $T^{\lambda\nu}_{a,\lambda} = 0$, then the expression $\mathcal{N}'^{\lambda}_a = \mathcal{N}^{\lambda}_a + T^{\lambda\nu}_{a,\nu}$ gives the same zero divergence property

$$\mathcal{N}'^{\lambda}_{a,\lambda} = \mathcal{N}^{\lambda}_{a,\lambda} + T^{\lambda\nu}_{a,\lambda} = \mathcal{N}^{\lambda}_{a,\lambda} = 0$$

8.2 Noether's Theorem for Gauge Symmetries

In the cases of gauge symmetries, that is, transformations of the field variable which are not generated by coordinate transformations, leaving the Lagrangian invariant, we also obtain conserved quantities from Noether's theorem, provided the symmetries are of the Lie type.

As we have seen in previous examples there are two major types of gauge transformations: global when the parameters of the group do not depend on the coordinates and local when the parameters depend on the coordinates. For each of these cases it is also usual to separate the gauge transformations in two kinds, depending on how the group acts:

$$\Psi' = \Psi + \xi(x) \tag{8.14}$$

$$\Psi' = e^{i\chi(x)}\Psi \tag{8.15}$$

respectively called gauge transformations of the first and the second kind. The first kind is like the gauge transformation in the electromagnetic theory. The second kind (also called a phase transformation) is like the unitary transformations in quantum mechanics.

Using the Lie algebra structure these two kinds of gauge transformations can be resumed in just one form when infinitesimal transformations are considered. Indeed, in the first kind (8.14) ξ has the same algebraic structure as the field Ψ. Therefore we may write it in terms of the field basis of the field space. Actually, since the field Ψ is also written in the same basis we may write ξ as $\xi = i\zeta(\theta, x)\Psi$, where ζ is an analytic complex function of the parameter θ. For an infinitesimal $\delta\theta$ we may expand this function in θ, keeping only the linear terms $\xi = \sum a_a \delta\theta^a \Psi$. The analyticity allows us to rebuild the finite transformations as in all Lie groups. Therefore the infinitesimal transformations of the first kind can be written as $\Psi' = \Psi + \sum a_a \delta\theta^a \Psi$. On the other hand, in the transformation of the second kind (8.15), χ is a scalar function which depends also analytically on θ. Therefore, we may expand the exponential and χ, keeping only the linear term in θ, obtaining $\chi = \sum a_a(x)\delta\theta^a$. Summarizing, for infinitesimal transformations of Lie groups, the two kinds of gauge transformations may be written as

$$\Psi' = \Psi + \sum a_a(x)\delta\theta^a\Psi \tag{8.16}$$

From its own nature, the gauge transformations do not produce a functional variation of the fields. Consequently, the total variation is the same as the algebraic variation. Using the notation for the total variation we may write

$$\delta_T\Psi = \sum_a \mathscr{F}_a(x, \Psi)\delta\theta^a \tag{8.17}$$

Notice that contrary to the case of coordinate transformations, the gauge transformation represents a map from a fiber to the same fiber as indicated in Fig. 8.3.

Theorem 8.2 (Noether's Theorem for Global Gauge Symmetry) *Given a field Ψ defined by a Lagrangian $\mathscr{L}(\Psi)$, invariant under a local Lie gauge symmetry G, there are N conserved quantities given by*

$$\mathscr{N}_a^\lambda = \frac{\partial \mathscr{L}}{\partial \Psi_{,\lambda}}\mathscr{F}_a(\Psi), \quad a = 1, \ldots, M$$

Indeed, in a global gauge transformation the parameters do not depend on the coordinates. Using the same notation of the previous theorem, consider the total variation of the Lagrangian $\delta_T\mathscr{L} = \mathscr{L}(\Psi'(x)) - \mathscr{L}(\Psi(x))$. Therefore the total variation of the action is

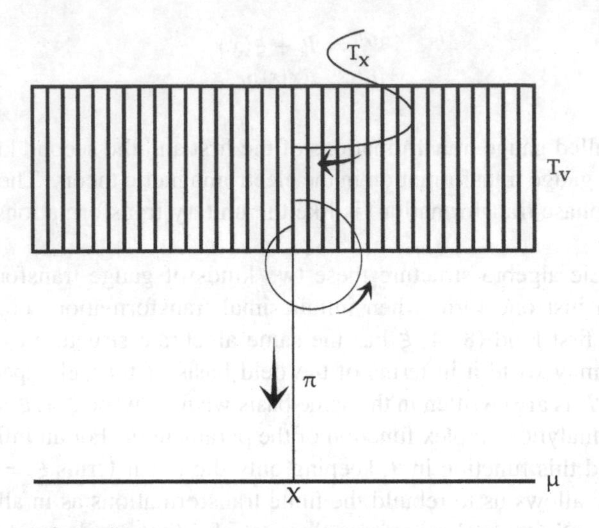

Fig. 8.3 Gauge transformation

$$\delta_T A = A(\Psi', \Omega) - A(\Psi, \Omega)$$

Unlike the case of coordinate transformations the region Ω does not change. Therefore

$$\delta_T A = \int_\Omega [\mathscr{L}(\Psi'(x)) - \mathscr{L}(\Psi(x))] dv = \int_\Omega \delta_T \mathscr{L} \, dv$$

However,

$$\delta_T \mathscr{L} = \frac{\partial \mathscr{L}}{\partial \Psi} \delta_T \Psi + \frac{\partial \mathscr{L}}{\partial \Psi_{,\lambda}} \delta_T \Psi_{,\lambda}$$

where $\delta_T \Psi$ and $\delta_T \Psi_{,\lambda}$ are given by (8.17) and its derivative. Therefore, the total variation of the action is

$$\delta_T A = \int_\Omega \left(\frac{\partial \mathscr{L}}{\partial \Psi} \delta_T \Psi + \frac{\partial \mathscr{L}}{\partial \Psi_{,\lambda}} \delta_T \Psi_{,\lambda} \right) dv \tag{8.18}$$

Assuming that the gauge transformation is a symmetry of the system and using the Euler–Lagrange equations

$$\frac{\partial \mathscr{L}}{\partial \Psi} = \frac{\partial}{\partial x^\lambda} \left(\frac{\partial \mathscr{L}}{\partial \Psi_{,\lambda}} \right)$$

we obtain (remembering that here $\delta\theta^a$ does not depend on x^i)

$$\delta_{\mathrm{T}} A = \delta\theta^a \int_\Omega \frac{\partial}{\partial x^\lambda}\left(\frac{\partial \mathscr{L}}{\partial \Psi_{,\lambda}} \mathscr{F}_a\right) dv = 0$$

or denoting the Noether quantity

$$\mathscr{N}_a^\lambda = \frac{\partial \mathscr{L}}{\partial \Psi_{,\lambda}} \mathscr{F}_a \tag{8.19}$$

For a closed region where $\mathscr{L}$ and Ψ and the derivatives $\Psi_{,\lambda}$ are continuous, we obtain the conserved quantities

$$\mathscr{N}_{a,\mu}^\mu = 0$$

As in the previous theorem, we may apply the divergence theorem to obtain the conserved quantities (8.19).

Theorem 8.3 (Noether's Theorem for Local Gauge Symmetry) *Given a field Ψ by a Lagrangian $\mathscr{L}(\Psi)$ invariant under a local Lie gauge symmetry G, there are N quantities*

$$\mathscr{N}^\lambda{}_a = \frac{\partial \mathscr{L}}{\partial \Psi_{,\lambda}} \mathscr{F}_a$$

which are conserved in the sense that $D_\lambda \mathscr{N}_a^\lambda = 0$, where $D_\lambda = I\partial_{,\lambda} + A_\lambda$, are matrices defined in the Lie algebra of G and where the matrix A_λ has entries $A^a{}_{\lambda b}$ defined by Noether's condition

$$\sum \mathscr{F}_a \frac{\partial \delta\theta^a}{\partial x^\lambda} = \sum_a \mathscr{F}_a A^a{}_{\lambda b} \delta\theta^b$$

As in the last case we have the same total variation

$$\delta_{\mathrm{T}} A = \int \left(\frac{\partial}{\partial x^\lambda}\left(\frac{\partial \mathscr{L}}{\partial \Psi_{,\lambda}}\right)\delta_{\mathrm{T}}\Psi + \frac{\partial \mathscr{L}}{\partial \Psi_{,\lambda}} \delta_{\mathrm{T}}\Psi_{,\lambda}\right) dv$$

Using the expressions $\delta_{\mathrm{T}}\Psi_{,\lambda} = (\partial/\partial x^\lambda)\delta_{\mathrm{T}}\Psi$ and $\delta_{\mathrm{T}}\Psi = \mathscr{F}_a(x, \Psi)\theta^a$, after applying the Euler–Lagrange equations, we obtain

$$\delta_{\mathrm{T}} A = \int_\Omega \frac{\partial}{\partial x^\lambda}\left(\frac{\partial \mathscr{L}}{\partial \Psi_{,\lambda}} \delta_{\mathrm{T}}\Psi\right) dv = \int_\Omega \frac{\partial}{\partial x^\lambda}\left(\frac{\partial \mathscr{L}}{\partial \Psi_{,\lambda}} \mathscr{F}_a \delta\theta^a\right) dv = 0$$

In contrast with the previous case, here we cannot remove $\delta\theta^a$ from the bracket because $\delta\theta^a$ depends on the coordinate x^μ. However, expanding the indicated derivatives we find

$$\int_{\Omega}\left[\frac{\partial}{\partial x^{\lambda}}\left(\frac{\partial\mathcal{L}}{\partial\Psi_{,\lambda}}\mathcal{F}_a\right)\delta\theta^a+\frac{\partial\mathcal{L}}{\partial\Psi_{,\lambda}}\mathcal{F}_a\frac{\partial\delta\theta^a}{\partial x^{\lambda}}\right]dv=0 \tag{8.20}$$

As we see, the term between square brackets is not a divergent. Noether avoided this difficulty by adding and subtracting the term

$$\sum_a \frac{\partial\mathcal{L}}{\partial\Psi_{,\lambda}}\mathcal{F}_a A^a_{\lambda\,b}\delta\theta^b$$

and imposing that $A^a_{\lambda\,b}$ satisfy the "Noether condition" (this is not a commonly used designation but just a reference name for later use)

$$\sum_a \mathcal{F}_a\frac{\partial\delta\theta^a}{\partial x^{\lambda}}-\sum_a \mathcal{F}_a A^a_{\lambda\,b}\delta\theta^b=0 \tag{8.21}$$

Noting that the parameter indices are the same as that of the Lie algebra we may anticipate that these conditions define the components $A^a_{\lambda\,b}$ of a vector-matrix A_λ called the gauge connection or the gauge field, defined in the Lie algebra of G. We shall see this in more detail in the next sections.

To find the conserved quantities, add and subtract the above-mentioned terms to (8.20), obtaining

$$\int\left[\frac{\partial}{\partial x^{\lambda}}\left(\frac{\partial\mathcal{L}}{\partial\Psi_{,\lambda}}\mathcal{F}_a\right)\delta\theta^a+\frac{\partial\mathcal{L}}{\partial\Psi_{,\lambda}}\mathcal{F}_a\frac{\partial\delta\theta^a}{\partial x^{\lambda}}+\frac{\partial\mathcal{L}}{\partial\Psi_{,\lambda}}\mathcal{F}_a A^a_{\lambda n}\delta\theta^b-\frac{\partial\mathcal{L}}{\partial\Psi_{,\lambda}}\mathcal{F}_a A^a_{\lambda n}\delta\theta^b\right]dv=0$$

and applying (8.21), it follows that

$$\int_{\Omega}\frac{\partial}{\partial x^{\lambda}}\left(\frac{\partial\mathcal{L}}{\partial\Psi_{,\lambda}}\mathcal{F}_a+\frac{\partial\mathcal{L}}{\partial\Psi_{,\lambda}}\mathcal{F}_b A^b_{\lambda\,a}\right)\delta\theta^a dv=0$$

In this way we have managed to get $\delta\theta^a$ out of the derivative and the above integral can be written as

$$\int\left[\left(\frac{\partial}{\partial x^{\lambda}}\delta^a_b+A^a_{\lambda b}\right)\left(\frac{\partial\mathcal{L}}{\partial\Psi_{,\lambda}}\mathcal{F}_b\right)\right]\delta\theta^b dv=0$$

or after denoting the vector-matrix derivative operator

$$D^a_{\lambda b}=\delta^a_b\partial_{,\lambda}+A^a_{\lambda b}$$

we may write

$$\int_{\Omega}\left[D^a_{\lambda b}\left(\frac{\partial\mathcal{L}}{\partial\Psi_{,\lambda}}\mathcal{F}_a\right)\right]\delta\theta^b dv=0$$

Using the notation

$$\mathscr{N}^{\lambda}{}_{a} = \frac{\partial \mathscr{L}}{\partial \Psi_{,\lambda}} \mathscr{F}_{a}$$

it follows that

$$\int_{\Omega} \left(D^{a}_{\lambda\, b} \mathscr{N}^{\lambda}{}_{a}\right) \delta\theta^{b} dv = 0$$

Assuming that the integrand is differentiable and that the integration region Ω is closed, we obtain

$$\left(D^{a}_{\lambda\, b} \mathscr{N}^{\lambda}{}_{a}\right) \delta\theta^{b} = 0$$

However, $\delta\theta^{a}$ are linearly independent parameters so that

$$D_{\lambda}{}^{a}_{b} \mathscr{N}^{\lambda}{}_{a} = 0 \tag{8.22}$$

Using a matrix notation, looking at $\mathscr{N}^{\lambda}$ as a column vector, we may write this expression as

$$D_{\lambda} \mathscr{N}^{\lambda} = 0, \quad D_{\lambda} = I\partial_{\lambda} + A_{\lambda}$$

which is a generalization of the divergence of $\mathscr{N}^{\lambda}$, with respect to the generalized derivative D_{λ}.

The relevant fact to be noted here is that $A^{a}_{\lambda\, b}$ defined by the above expression is a vector-matrix field with respect to the space–time index λ, with matrix indices a, b belonging to the Lie algebra of the gauge group.

We shall see that A_{λ} is a connection associated with that adjoint representation of the Lie symmetry group.

Remark. For simplicity the above results were obtained by considering a real field Ψ. However, to keep up with the unitary gauge transformations the field becomes complex. Thus, depending on its construction the Lagrangian must also contain the complex conjugate fields. For example, for a conserved complex scalar field φ, of the complex Klein–Gordon field (6.10), the Lagrangian depends also on φ^*

$$\mathscr{L} = \mathscr{L}(\varphi, \varphi^*, \varphi_{,\lambda}, \varphi^*_{,\lambda}) \tag{8.23}$$

In these cases we must calculate the variations of the Lagrangian with respect to the field and of its complex conjugate in such a way that the conserved quantities $\mathscr{N}_{a}^{\lambda}$, which are observables, must be real:

$$\mathscr{N}_a^\lambda = \frac{\partial \mathscr{L}}{\partial \varphi_{,\lambda}} \mathscr{F}_a(\varphi) + \frac{\partial \mathscr{L}}{\partial \varphi^*_{,\lambda}} \mathscr{F}_a(\varphi^*)$$

For complex matrix fields the situation is slightly more complicated because we need to consider the Hermitian conjugate field $\Psi^\dagger$, instead of the complex conjugate. Since the resulting expressions for the Hermitian conjugate are similar to those of the field itself it is usual just to add to the conserved quantity just the symbol +HC to indicate "plus the Hermitian conjugate expression."

Let us detail the above remark for the case of the local gauge transformation of (8.23).

For an infinitesimal transformation of the unitary group $\varphi'(x) = u(x)\varphi'(x) = e^{i\theta(x)}\varphi, \;\; u^*u = 1$, the total variation of φ and of its derivative is

$$\delta_T \varphi = i\theta\varphi \;\text{ and }\; \delta_T \varphi_{,\lambda} = i(\theta\varphi_{,\lambda} + \varphi\theta_{,\lambda}) \tag{8.24}$$

and similarly for φ^*. Then the total variation of the Lagrangian is

$$\delta_T \mathscr{L} = \frac{\partial \mathscr{L}}{\partial \varphi}\delta_T\varphi + \frac{\partial \mathscr{L}}{\partial \varphi_{,\lambda}}\delta_T\varphi_{,\lambda} + \frac{\partial \mathscr{L}}{\partial \varphi^*}\delta_T\varphi^* + \frac{\partial \mathscr{L}}{\partial \varphi^*_{,\lambda}}\delta_T\varphi^*_{,\lambda}$$

or using the notation +HC also for the complex conjugate we may write

$$\delta_T \mathscr{L} = i\theta\left(\frac{\partial \mathscr{L}}{\partial \varphi}\delta_T\varphi + \frac{\partial \mathscr{L}}{\partial \varphi_{,\lambda}}\delta_T\varphi_{,\lambda}\right) + i\theta_{,\lambda}\frac{\partial \mathscr{L}}{\partial \varphi_{,\lambda}}\varphi + \text{HC}$$

Applying the Euler–Lagrange equations

$$\begin{aligned}
\delta_T \mathscr{L} &= i\delta\theta\left(\frac{\partial}{\partial x^\lambda}\left(\frac{\partial \mathscr{L}}{\partial \varphi_{,\lambda}}\right)\delta_T\varphi + \frac{\partial \mathscr{L}}{\partial \varphi_{,\lambda}}\delta_T\varphi_{,\lambda}\right) + i\delta\theta_{,\lambda}\frac{\partial \mathscr{L}}{\partial \varphi_{,\lambda}} + \text{HC} \\
&= i\delta\theta\frac{\partial}{\partial x^\lambda}\left(\frac{\partial \mathscr{L}}{\partial \varphi_{,\lambda}}\delta_T\varphi\right) + i\delta\theta_{,\lambda}\left(\frac{\partial \mathscr{L}}{\partial \varphi_{,\lambda}}\delta_T\varphi\right) + \text{HC} \\
&= i\delta\theta\left[\partial_\lambda + \frac{\delta\theta_{,\lambda}}{\delta\theta}\right]\left(\frac{\partial \mathscr{L}}{\partial \varphi_{,\lambda}}\delta_T\varphi\right) + \text{HC}
\end{aligned} \tag{8.25}$$

showing that it is the existence of the derivative of the parameter that prevents the emergence of the divergence term. Noether's theorem for local gauge symmetry tells that this can be fixed by changing the partial derivative $\partial_\lambda\varphi$ to the *gauge covariant derivative*

$$D_\lambda\varphi = (\partial_\lambda + iA_\lambda)\varphi$$

where A_λ is a 1×1 matrix-vector field such that its components satisfy Noether's condition

$$\mathscr{F}\frac{\partial\delta\theta}{\partial x^{\lambda}} = \mathscr{F}A_{\lambda}\delta\theta$$

Here in this simple case we may cancel the function $\mathscr{F}$, yielding to the simpler solution

$$A_{\lambda} = \frac{\delta\theta_{,\lambda}}{\delta\theta}$$

which is consistent with the vanishing of $\delta_{T}\mathscr{L}$ in (8.25).

Therefore, the covariant derivative becomes $D_{\lambda} = \partial_{\lambda} + i\theta_{,\lambda}/\theta$ and the invariant Lagrangian must be rewritten in terms of this derivative as

$$\mathscr{L} = \mathscr{L}(\varphi, D_{\lambda}\varphi, \varphi^{*}, (D_{\lambda}\varphi)^{*})$$

In this case the Noether quantity is just a vector (a 1×1 matrix) with components

$$\mathscr{N}^{\lambda} = \frac{\partial\mathscr{L}}{\partial\varphi_{,\lambda}}\varphi + \mathrm{HC}$$

which is conserved in the sense that $D_{\lambda}\mathscr{N}^{\lambda} = 0$.

An interesting question is: How does a gauge field A transform under a local gauge transformation?

Theorem 8.4 (Gauge Transformation of A) *Given a field Ψ and a local gauge transformation, $\Psi' = u\Psi$, the gauge field transforms as*

$$A'_{\mu} = uA_{\mu}u^{-1} + u_{,\mu}u^{-1}$$

Indeed, consider the gauge transformation acting as an operator in the space of the field $\Psi' = u\Psi$. The transformation of the gauge covariant derivative of $D_{\mu}\Psi = (I\partial_{\mu} + A_{\mu})\Psi$ is

$$(D_{\mu}\Psi)' = ((\partial_{\mu} + A_{\mu})\Psi)' = (\partial_{\mu} + A'_{\mu})u\Psi = \partial_{\mu}(u\Psi) + A'_{\mu}u\Psi$$

On the other hand, u also acts as an operator on the derivative of the field $(D_{\mu}\Psi)' = u(D_{\mu}\Psi)$. Comparing this with the above expression we obtain

$$u(D_{\mu}\Psi) = \partial_{\mu}(u\Psi) + A'_{\mu}u\Psi$$

or,

$$u(I\partial_{\mu} + A_{\mu})\Psi = u_{,\mu}\Psi + u\partial_{\mu}\Psi + A'_{\mu}u\Psi$$

Multiplying this expression at left by u^{-1} and at right by u, gives after canceling Ψ

$$A'_\mu = u A_\mu u^{-1} + u_{,\mu} u^{-1} \tag{8.26}$$

Theorem 8.5 (Noether's Theorem in General Relativity) *Given an infinitesimal coordinate transformation in a space–time of general relativity and a field Ψ minimally coupled to the gravitational field, then the energy–momentum tensor of Ψ is conserved as long as there exists a Killing vector field in the space–time.*

One fundamental property of general relativity is that it is invariant under the diffeomorphism group of the space–time (the diffeomorphism invariance of the theory). The minimal coupling of a field Ψ with gravitation represents the simplest way to implement the general covariance to Ψ. In such procedure we just replace the partial derivatives by the covariant derivative, without adding any new terms proportional to the Riemann tensor and its contractions.

Consider an infinitesimal coordinate transformation in an arbitrary space–time solution of Einstein's equations

$$x'^\alpha = x^\alpha + \xi^\alpha$$

where the descriptor ξ is a differentiable function of the coordinates and an analytic function of the parameters. Expanding ξ^α gives

$$\xi^\alpha = \sum_{m=0}^{N} a^\alpha_a \theta^a$$

The action integral for the field Ψ minimally coupled to the gravitational field $g_{\mu\nu}$ is then written as

$$A = \int_\Omega \mathscr{L}(\Psi, \Psi_{;\mu}) dv$$

which differs from the Minkowski field theory in that now we have the covariant derivative defined by the metric connection. Therefore, the total variation of the action resulting from the total variation of Ψ and its covariant derivative is

$$\delta_{\mathrm{T}} A = \int_\Omega \left(\frac{\partial \mathscr{L}}{\partial \Psi_{;\lambda}} \delta_{\mathrm{T}} \Psi + \mathscr{L} \xi^\lambda \right)_{;\lambda} dv$$

Recalling the definition of the total variation of a field given by (8.9), it is dependent on the field and its (partial) derivatives: $\delta_{\mathrm{T}} \Psi = \sum (\mathscr{F}_a - \Psi_{,\lambda} a^\lambda_a) \delta\theta^a$. We may simplify this expression by defining a new functional $\tilde{\mathscr{F}}_\lambda a^\lambda_a = \mathscr{F}_a$, so that

$$\delta_{\mathrm{T}} \Psi = (\tilde{\mathscr{F}}_\lambda - \Psi_{,\lambda}) a^\lambda_a \theta^a$$

Denoting $\mathscr{U}_\lambda = (\tilde{\mathscr{F}}_\lambda - \mathscr{F}_\lambda)$ we may finally rewrite the total variation of Ψ as

$$\delta_T \Psi = \mathscr{U}_\lambda \xi^\lambda$$

and the total variation of the action is

$$\delta_T A = \int_\Omega \left(\frac{\partial \mathscr{L}}{\partial \Psi_{;\lambda}} \mathscr{U}_\mu \xi^\mu + \mathscr{L} \xi^\lambda \right)_{;\lambda} dv = \int_\Omega \left[\left(\frac{\partial \mathscr{L}}{\partial \Psi_{;\lambda}} \mathscr{U}_\mu + \mathscr{L} \delta^\lambda_\mu \right) \xi^\mu \right]_{;\lambda} dv = 0$$

Denoting the symmetric energy-momentum tensor of the field by $T_{\mu\nu} = g\lambda\nu T^\lambda_\mu$ with

$$T^\lambda_\mu = \frac{\partial \mathscr{L}}{\Psi_{;\lambda}} \mathscr{U}_\mu + \mathscr{L} \delta^\lambda_\mu$$

and defining the Noether quantity $\mathscr{N}^\lambda = T^\lambda{}_\mu \xi^\mu$, we obtain

$$\int_\Omega (T^\lambda{}_\mu \xi^\mu)_{;\lambda} dv = \int_\Omega \mathscr{N}^\lambda{}_{;\lambda} dv = 0 \tag{8.27}$$

which is equivalent to the Noether theorem for local gauge (8.22). The difference is that here we have a coordinate transformation of the group diffeomorphisms of the space–time manifold (which is an infinite Lie group), and the connection is introduced independently by a postulate.

Nonetheless, assuming that the integrand in (8.27) is differentiable, it is possible to determine conserved quantities for a closed region of integration. Calculating the covariant derivative we find that

$$\mathscr{N}^\lambda{}_{;\lambda} = T^\lambda{}_{\mu;\lambda} \xi^\mu + T^{\lambda\mu} \xi_{\mu;\lambda} = 0$$

Since $T^{\mu\nu} = g^{\mu\lambda} T^\nu_\lambda$ is a symmetric tensor, the last term in the above expression is equivalent to $\xi_{(\mu;\lambda)}$. Therefore, the quantity

$$T^\lambda{}_{\mu;\lambda} = 0$$

is conserved in the sense of Noether's theorem only if we have

$$\xi_{(\mu;\lambda)} = 0$$

This is called the Killing equation defining an infinitesimal isometric coordinate transformation in the space–time and ξ is a Killing vector field.

Since not all solutions of Einstein's equations admit Killing vector fields, actually only a very small class of solutions has such isometry groups, the existence of conserved quantities in the sense of Noether is very restricted in general relativity.

Chapter 9
Bundles and Connections

In previous chapters we have described how the notion of symmetry of a physical theory forced us to change the definition of derivative of a field so as to make the action principle invariant. The invariance of the action means that it must be the same for all observers (when considering coordinate transformations) and for the observable fields (when referring to gauge transformations). It has to be such invariant to fulfill the purposes of Maupertuis, Euler, Lagrange, and Hamilton. The modification of the derivative is quite intuitive when we are talking about coordinate transformations, but it is less intuitive when we are talking about field theory. In this chapter we will introduce the basic tools to the formal definition of gauge covariant derivatives.

Starting with the more intuitive coordinate transformations, we have a very naive notion of Euclidean 3-dimensional background Euclidean space where we make frequent use of Cartesian coordinates defined in an absolute (or canonical) reference frame. The necessity to change coordinates away from the Cartesian system was regarded as a way to simplify the equations and not as a fundamental issue. However, as we have already exemplified in the previous chapters, in doing mathematical analysis in a physical manifold we no longer have the Cartesian frames at our disposal. In order to make sense as an invariant property to the allowed observers, the notion of derivative needs to be modified by the introduction of a connection or covariant derivative. In the following we generalize the notions of vector bundles to *principal fiber bundles*, so that we may incorporate a symmetry group to our physical manifold structure [108–112].

9.1 Fiber Bundles

Definition 9.1 (Fiber Bundle) Consider an n-dimensional differentiable manifold $\mathscr{M}$, to each point p of which there is an attached another m-dimensional manifold $\mathscr{B}_p$. The word attachment means that in principle there is not any specific relationship between $\mathscr{M}$ and $\mathscr{B}_p$, besides the point of contact. Then we may collect all these manifolds $\mathscr{B}_p$ in a set $T\mathscr{B}$, in such a way that we may identify the contact point p in $\mathscr{M}$ by a map $\pi : T\mathscr{B} \to \mathscr{M}$. The differentiable *fiber bundle* with base $\mathscr{M}$ and total space $T\mathscr{B}$ is the triad

M.D. Maia, *Geometry of the Fundamental Interactions*,
DOI 10.1007/978-1-4419-8273-5_9,

$$(\mathscr{M},\ \pi,\ T\mathscr{B})$$

If $p \in \mathscr{M}$, then the map $s(p) \rightarrow T\mathscr{B}$ such that $s(p) \circ \pi = 1$ is called a section of the fiber bundle, all it does is to identify the fiber $\mathscr{B}_p \subset T\mathscr{B}$.

The tangent bundle is a particular fiber bundle. Higher order tangent bundles such as the osculating paraboloid to a surface are interesting non-trivial examples.

Other examples are given by Galilean and Newton's space–time in which the total space is the space–times and the fibers are of the three-dimensional simultaneity sections. In both cases the base space is the absolute time axis $I\!R$.

Given a fiber bundle $(\mathscr{M},\ \pi,\ T\mathscr{B})$ and neighborhood of a point x, $\cup_x \subset \mathscr{M}$, we may define a *local fiber bundle* as the restriction of a fiber bundle to the points of $\cup_x$

$$\text{Local fiber bundle} = (\cup_x,\ \pi,\ T\mathscr{B}\rfloor_{\cup_x})$$

As interesting examples consider the cylinder (Fig. 9.1) and Möbius strip (Fig. 9.2) as fiber bundles.

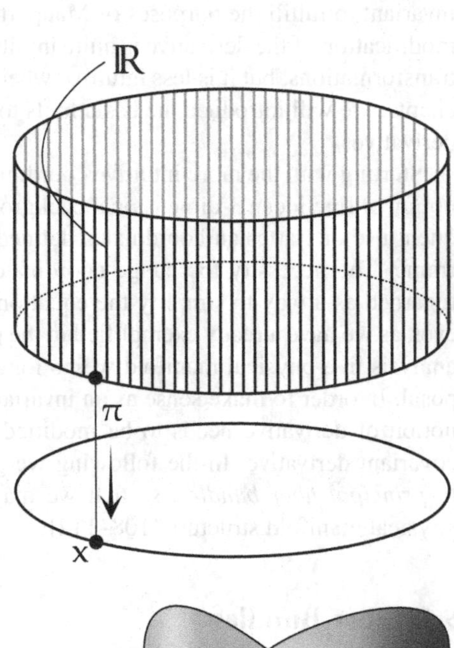

Fig. 9.1 Cylinder fiber bundle

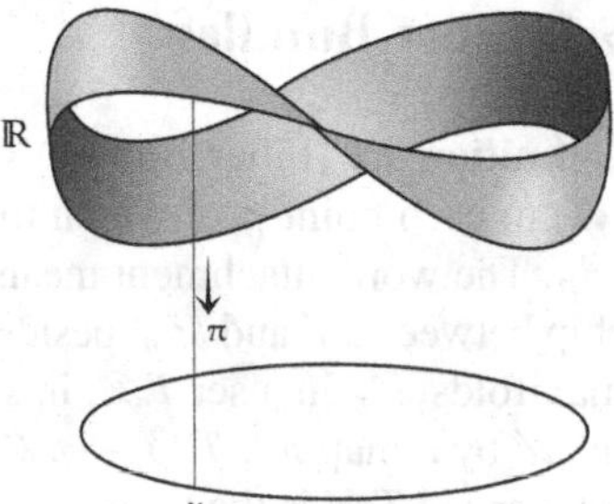

Fig. 9.2 The Möbius fiber bundle

The cylinder can be described as a fiber bundle with a circle S^1 as the base space and where each fiber is a segment of line (a compact manifold) with fixed length ℓ. The total space is then the set of lines in the rectangle $I\!R^2$, with identified lateral sides forming a cylinder, where the fibers are the line segments at each point, and the total space is the cylinder itself

$$(S^1, \pi, S^1 \times \ell)$$

On the other hand, by giving a rotation of each fiber around the middle line, we obtain as the total space the Möbius strip itself

$$(S^1, \pi, \text{ Möbius})$$

The helicity of the fibers prevents the identification of the total space Möbius with $S^1 \times \ell$. However, locally they can be identified.

9.2 Base Morphisms

Consider two fiber bundles $(\mathscr{M}, \pi, T\mathscr{B})$ and $(\mathscr{M}', \pi, T\mathscr{B}')$. A *morphism* between them is a differentiable map

$$\varphi : T\mathscr{B} \to T\mathscr{B}'$$

such that it takes a fiber of $T\mathscr{B}$ into a fiber of $T\mathscr{B}'$. In particular, two differentiable fiber bundles are isomorphic when φ is 1:1. If we take the bases $\mathscr{M}$ and $\mathscr{M}'$ to be the same, then we have a *base-isomorphism.*

Definition 9.2 (Trivial Fiber Bundle) A fiber bundle with base $\mathscr{M}$ and fibers $\mathscr{B}_p$ is said to be a *trivial fiber bundle* when the total space is the Cartesian product of the base $\mathscr{M}$ with a manifold Σ that is isomorphic to all fibers $\mathscr{B}_p$

$$(\mathscr{M}, \pi, \mathscr{M} \times \Sigma)$$

Σ is called the *typical fiber*.

The designation trivial fiber bundle results from the fact that the total space can be represented by a box where its elements (x, v) with $x \in \mathscr{M}$ and $v \in \Sigma$ are the sides of the box: In a trivial fiber bundle the fiber over p is

$$\mathscr{B}_p = (p, \Sigma) = \{(p, y), \ \forall\, y \in \Sigma\}$$

A simple example of trivial fiber bundle is the cylinder bundle shown in Fig. 9.1. In the particular case where the typical fiber Σ is a vector space we have a *trivial vector bundle*.

Definition 9.3 (Trivialization of Fiber Bundles) A fiber bundle is said to be trivializable when there is a base morphism

$$(\mathcal{M},\ \pi,\ T\mathcal{B}) \to (\mathcal{M},\ \pi,\ \mathcal{M} \times \Sigma)$$

Note from this definition that for a given fiber bundle we may have several different trivializations, even considering those with the same typical fiber Σ, but with different base morphisms.

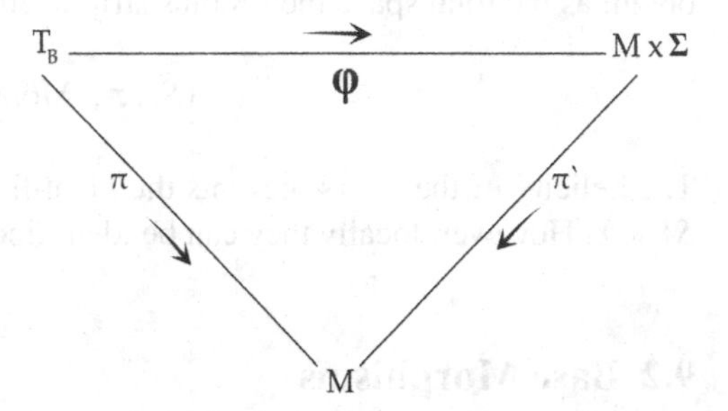

Fig. 9.3 Trivialization of a fiber bundle

Example 9.1 (Galilean Space–Time) As we have seen, the Galilean space–time $\mathcal{G}_4$ is the total space of a fiber bundle where the fibers are the simultaneity sections Σ_t and the base space is the absolute time axis $I\!R$.

$$(I\!R,\ \pi,\ \mathcal{G}_4)$$

In that space–time each simultaneous sections Σ_t is isomorphic to $I\!R^3$, so that $\mathcal{G}_4 = I\!R \times I\!R^3$. Therefore this fiber bundle is trivialized by the existence and the properties of the absolute time:

$$(I\!R,\ \pi,\ \mathcal{G}_4) \longrightarrow (I\!R,\ \pi,\ I\!R \times I\!R^3)$$

Since $\mathcal{G}_4$ is parameterized by $I\!R^4$, it follows that the trivialization can be defined by a specific choice of coordinate chart of $\mathcal{G}_4$ which are compatible with the Galilean group. For each Galilean transformation we have a trivialization, with the same typical fiber $I\!R^3$.

In general when a symmetry is not specified then the coordinates are mapped onto one another by the diffeomorphism group of the manifold. In this case, two trivializations of the same fiber bundle are said to be compatible when there is a diffeomorphism between the two typical fibers. Given a fiber bundle $(\mathcal{M},\ \pi,\ T\mathcal{B})$, and two trivializations of it given by

$$\phi : T\mathcal{B} \to \mathcal{M} \times \Sigma \quad \text{and} \quad \phi' : T\mathcal{B} \to \mathcal{M} \times \Sigma$$

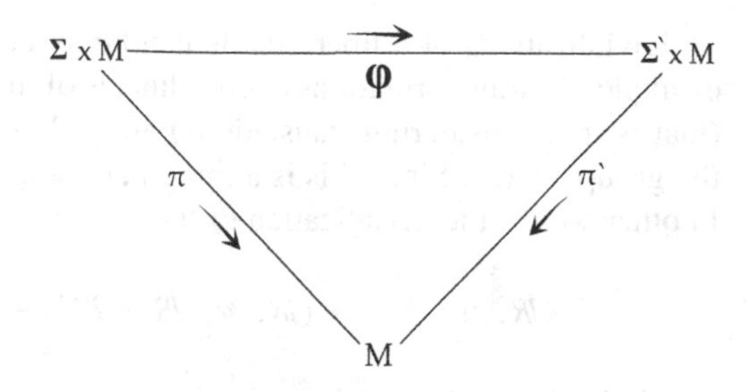

Fig. 9.4 Compatible trivializations

then the two trivializations are said to be equivalent when there is a base morphism φ such that the typical fiber of one is mapped in the typical fiber of the other, that is,

$$\varphi : \Sigma \to \Sigma'$$

Since $\pi = \pi' \circ \varphi$ we obtain

$$\varphi = \pi'^{-1} \circ \pi$$

which is the compatibility condition for equivalence. Two trivializations of the same fiber bundle are said to be equivalent when there is a base morphism between them.

Definition 9.4 (Local Trivialization) A given fiber bundle $(\mathscr{M}, \pi, T\mathscr{B})$ is *locally trivializable* when

a) $\forall x \in \mathscr{M}$, there is a neighborhood U_x and manifold Σ such that

$$T\mathscr{B}\rfloor_{\mathsf{U}_x} = \mathsf{U}_x \times \Sigma$$

b) The manifold Σ is diffeomorphic to all fibers.

Example 9.2 (The Möbius Strip) An example of locally trivial fiber bundle is given by the Möbius strip. All fibers are equal segments of lines but they have different orientations (a helicity), which are diffeomorphic to a single segment.

Example 9.3 (The Newtonian Space–Time) Another example is given by Newton's space–time $\mathscr{N}_4$ associated with the fiber bundle

$$(I\!R,\ \pi, \mathscr{N}_4)$$

As in the Galilean case, the base of thc fiber bundle is the absolute time axis. However, the simultaneity sections are only locally defined in the neighborhood U_t of the time interval, isomorphic to $I\!R^3$. Therefore, Newton's space–time is *only locally trivialized* by the absolute time and its properties

$$(I\!R,\ \pi, \mathscr{N}_4) \to (\mathsf{U}_t,\ \pi, \mathsf{U}_t \times I\!R^3)$$

Trivializations of a fiber bundle can be associated with a symmetry group. For example, the transformations of coordinates of the Galilean group with a fixed origin (that is, not considering translations) are orthogonal transformations belonging to the group $SO(3)$. Since this is a three-parameter Lie group it is isomorphic to $I\!R^3$. In other words, the trivialization of the Galileo space–time can be written as

$$(I\!R,\ \pi,\ \mathscr{G}_4) \rightarrow (I\!R,\ \pi,\ I\!R \times I\!R^3) \rightarrow (I\!R,\ \pi,\ I\!R \times SO(3))$$

Similarly, for the Newtonian space–time, the generalized Galilean group

$$\begin{cases} x'^i = A^{\mu}_{j} x_j + c^i(t) \\ t' = at + b \\ \phi' = \phi + \dfrac{\partial^2 c^i(t)}{\partial t^2} x^i \end{cases}$$

and all simultaneity sections are locally isomorphic to $I\!R^3$. Fixing $c(x) = 0$ the group resumes to $S0(3)$ and we obtain the local trivialization with the same group

$$(I\!R, \pi,\ \mathscr{N}_4) \rightarrow (I\!R, \pi,\ \mathsf{U}_t \times I\!R^3) \rightarrow (\mathsf{U}_t,\ \pi,\ \mathsf{U}_t \times SO(3))$$

These particular examples suggest the emergence of another type of fiber bundle which is associated with a Lie symmetry group G, called the principal fiber bundle of G.

9.3 Principal Fiber Bundles

Definition 9.5 (Principal Fiber Bundle) The *principal fiber bundle* or simply the principal bundle of a Lie group G is a fiber bundle with base $\mathscr{M}$ and where G *acts on the total space* $\mathscr{B}$ *as a map between fibers.* To make explicit the presence of G, a principal fiber bundle is usually denoted by a ordered tetrad

$$(G,\ \mathscr{M},\ \pi,\ T\mathscr{B})$$

The requirement that G is a Lie group will become clear in the next chapter. For now, it is sufficient to remind the fact that for a Lie group we can always deal with infinitesimal transformations, leading to its Lie algebra $\mathscr{G}$. Therefore, the principal fiber bundle of a Lie group can be always written in terms of its Lie algebra as

$$(\mathscr{G},\ \mathscr{M},\ \pi,\ T\mathscr{B})$$

From now on we shall refer to principal fiber bundles using *only its Lie algebra.*

Trivializations of principal fiber bundles are defined as before, given by a base morphism $\varphi : T\mathscr{B} \to \Sigma \times \mathscr{M}$, with the additional condition that *the Lie algebra* $\mathscr{G}$ *acts linearly upon the typical fiber* Σ

$$\mathscr{G} : \Sigma \to \Sigma$$

leading to the trivial principal bundle

$$(\mathscr{G},\ \mathscr{M},\ \pi,\ \mathscr{M} \times \Sigma)$$

where, as we said, the operators of $\mathscr{G}$ act as linear operators on the typical fiber Σ.

A particularly interesting trivialization is that defined by the space of the Lie algebra itself: *A trivialization of the principal bundle induces a particular representation of the Lie algebra where the representation space, is the space of the Lie algebra.* This particular representation was defined in Chapter 3 as the *adjoint representation* of $\mathscr{G}$. As we recall, this representation is defined by the structure constants of the group.

Reciprocally, *the adjoint representation of the Lie algebra of a group G induces a trivialization of the principal fiber bundle of G*.

The adjoint representation of a Lie algebra was defined as operators acting on the space of the Lie algebra, acting on its basis $\{X_a\}$ as

$$\tilde{\mathscr{G}}(X_a)X_b \stackrel{\text{def}}{=} [X_a, X_b] = f^c{}_{ab}X_c$$

Therefore, the adjoint representation is *unique as it is completely determined by the structure constants of the group* $f^c{}_{ab}$. This uniqueness is relevant because as we remember, in general, representations of a group are arbitrarily chosen. This is not the case of the adjoint representation which is self-contained in the Lie algebra structure.

From now on, we will denote the adjoint representation of a Lie algebra $\mathscr{G}$ by $\tilde{\mathscr{G}}$. It is the same algebra whose space is acted upon by the group (or algebra). Therefore, the trivialization of the principal bundle of a Lie group G by the adjoint representation of its Lie algebra is

$$(\tilde{\mathscr{G}},\ \mathscr{M},\ \pi,\ \mathscr{M} \times \tilde{\mathscr{G}})$$

As we see there is a notational redundancy, where $\tilde{\mathscr{G}}$ shows repeatedly. It is no longer necessary to specify the Lie algebra twice, and it has become common practice to denote the trivialized principal fiber bundle by the adjoint representation simply by the usual ordered triad

$$(\mathscr{M},\ \pi,\ \mathscr{M} \times \tilde{\mathscr{G}})$$

Using the adjoint representation we may reexamine the two previous examples.

Example 9.4 (Trivialization of the Galilean Space–Time) Taking the Lie algebra $\mathcal{G}_{SO(3)}$ as a subalgebra of the Galilean group defined in the Galilean space–time, the principal fiber bundle of this group is

$$(\mathcal{G}_{SO(3),I\!R},), \ \pi, \ \mathcal{G}_4)$$

which is trivialized to

$$(\mathcal{G}_{SO(3),I\!R},), \ \pi, \ I\!R \times \Sigma)$$

where the typical fiber Σ is isomorphic to all simultaneity sections and isomorphic to $I\!R^3$, upon which the $\mathcal{G}_{SO(3)}$ group acts. In particular taking Σ to be the space of the Lie algebra $\tilde{\mathcal{G}}_{SO(3)}$, we obtain the adjoint representation $\tilde{\mathcal{G}}_{SO(3)}$ and therefore the trivialization.

9.4 Connections

The purpose of the trivialization of a principal bundle by the adjoint representation is to obtain a connection.

As we have seen, in the adjoint representation the Lie algebra acts on itself, inducing a trivialization of the principal fiber bundle of G with base $\mathcal{M}$ to

$$(\mathcal{M}, \ \pi, \ \mathcal{M} \times \tilde{\mathcal{G}})$$

In this trivial fiber bundle the fibers are all isomorphic to the Lie algebra space $\mathcal{G}$.

Now, suppose that our field Ψ defined on $\mathcal{M}$ by a Lagrangian $\mathcal{L}(\Psi, \Psi_{,\mu})$ has a Lie symmetry group G. This means that G and hence its Lie algebra $\mathcal{G}$ act on Ψ, keeping the Lagrangian invariant. Since in the adjoint representation $\tilde{\mathcal{G}}$ acts on the algebra space $\mathcal{G}$, then in this representation Ψ corresponds to an element of $\mathcal{G}$. Using this double role of the adjoint representation, we may express the field Ψ as linear combination of the Lie algebra basis $\{X_a\}$

$$\Psi = \sum \Psi^a X_a$$

or equivalently, the field can also be written in terms of the dual of the Lie algebra $\tilde{\mathcal{G}}^*$. This provides another equivalent trivialization of the principal fiber bundle of G, the *dual adjoint trivialization*

$$(\mathcal{M}, \ \pi, \ \mathcal{M} \times \tilde{\mathcal{G}}^*)$$

Using this representation, the same field Ψ is written in terms of the dual basis $\{X^a\}$ of the Lie algebra (defined by $X^a(X_b) = \delta^a_b$), so that Ψ is now regarded as a one-form field:

$$\Psi = \sum \Psi_a X^a$$

We may now consider the three types of symmetry:

(a) When G is a *coordinate symmetry*, such as the Poincaré group, we may express the coordinate transformation generally as

$$x'^{\mu} = f^{\mu}(x^{\nu}, \theta_a)$$

Then by its definition $\{X^a\}$ can be expressed directly in terms of the coordinate basis as (from (3.7))

$$X_a = \sum a^{\mu}_a(x)\frac{\partial}{\partial x^{\mu}}$$

and its dual

$$X^a = \sum a^a_{\mu}(x)dx^{\mu}$$

Therefore, in the dual adjoint representation of a coordinate transformation we may express the field as a one-form field

$$\Psi = \sum \Psi_a X^a = \sum \Psi_a a^a_{\mu}(x)dx^{\mu} = \sum \Psi_{\mu}dx^{\mu}$$

where we have denoted $\Psi_{\mu} = \Psi_a a^a_{\mu}(x)$. Therefore the exterior derivative of the field Ψ gives a two-form field

$$d \wedge \Psi = \sum d\Psi_{\mu} \wedge dx^{\mu} = \sum \frac{\partial \Psi_{\mu}}{\partial x^{\nu}} dx^{\nu} \wedge dx^{\mu}$$

(b) In the case of a *global gauge symmetry* of a field defined on a space–time, the gauge transformation is

$$\Psi'^{\mu} = f^{\mu}(\Psi, \theta)$$

where in the global case, θ does not depend on the coordinates of the space–time. Therefore, given the adjoint representation of the Lie algebra of the gauge group with generators X_a, by the same token we may express the field Ψ in terms of the Lie algebra basis $\{X_a\}$, or of its dual $\{X^a\}$ as

$$\Psi = \sum \Psi^a X_a = \sum \Psi_a X^a \tag{9.1}$$

Since in the global case X^a do not depend on the coordinates x^μ, the exterior derivative of Ψ is a two-form field given by

$$d \wedge \Psi = \sum d\Psi_a \wedge X^a$$

However, the components Ψ_a of the field are functions of the space–time coordinates, so that we may write $d\Psi_a = \Psi_{a,\mu} dx^\mu$. Consequently, as in the previous case, the exterior derivative of Ψ can also be written in terms of the dual basis of the tangent bundle of $\mathscr{M}$ as

$$d \wedge \Psi = \sum d\Psi_a \wedge X^a = -\sum \Psi_{a,\mu} X^a \wedge dx^\mu \tag{9.2}$$

(c) For *local gauge symmetries*, the group and its Lie algebra are locally defined. Therefore its base $\{X_a\}$ and the dual $\{X^a\}$ also depend on the coordinates of $\mathscr{M}$. In this case, the exterior derivative acts on both factors of the one-form field

$$\Psi = \sum \Psi_a X^a \tag{9.3}$$

as

$$d \wedge \Psi = \sum d\Psi_a \wedge X^a + \Psi_a \, d \wedge X^a \tag{9.4}$$

where $d \wedge X^a$ is a two-form. As such it can be expressed as an exterior product of X^a with another one-form ω^a_b belonging to the same space:

$$d \wedge X^a = \sum \omega^a{}_b \wedge X^b$$

Now, we have a more complicated situation as compared with (9.2), because we need to relate also the one-form X^b to the cotangent coordinate basis dx^μ (Fig. 9.5). This relation is formally done by the derivative map of a base morphism between the cotangent bundle $(\mathscr{M},\ \pi,\ T\mathscr{M}^*)$ and the trivialized dual principal fiber bundle $(\mathscr{M},\ \pi,\ \tilde{\mathscr{G}}^*)$ which is defined by the Jacobian matrix $\left(\dfrac{\partial f^a}{\partial x^\mu}\right)$ of the transformation between basis

$$X^a = \frac{\partial f^a}{\partial x^\mu} dx^\mu \tag{9.5}$$

With this transformation we may express

$$d \wedge X^a = \omega^a_b \wedge \frac{\partial f^b}{\partial x^\mu} dx^\mu = \omega^a_\mu \wedge dx^\mu$$

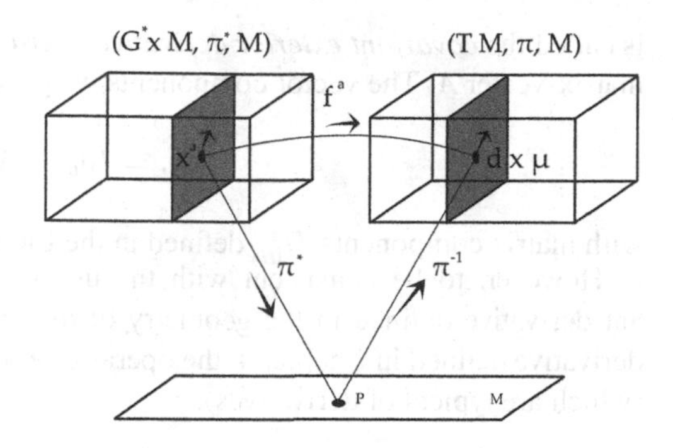

Fig. 9.5 Correspondence between X^a and dx^μ

where we have denoted

$$\omega^a_\mu = \omega^a_b \frac{\partial f^b}{\partial x^\mu}$$

Since ω^a_μ is a one-form with components in the dual adjoint representation $\tilde{\mathscr{G}}^*$, it may be expressed in terms of the dual basis $\{X^a\}$ as

$$\omega^a_\mu = -A^a_{b\,\mu} X^b$$

After inverting the order of the last exterior product and replacing in (9.4), we obtain

$$d \wedge \Psi = \Psi_{a,\mu} dx^\mu \wedge X^a - \Psi_a A^a_{\mu b} X^b \wedge = (\delta^a_b \partial_\mu + A^a_{\mu b})\, \Psi_a \, dx^\mu \wedge X^b$$

or, denoting

$$D^a_{\mu b} = \delta^a_b \partial_\mu + A^a_{\mu b} \tag{9.6}$$

we may write the exterior derivative of the field for a local gauge symmetry as

$$d \wedge \Psi = \sum D^a_{\mu b}\, \Psi_a \, dx^\mu \wedge X^b$$

The expression $D^a_{\mu b}\, \Psi_a$ extends the exterior derivative by the inclusion of the coefficients $A^a_{\mu b}$ defined in the Lie algebra of the local symmetry group.

Definition 9.6 (Gauge Connection) Given a Lie group G acting as symmetry of a physical field Ψ the vector-matrix derivative operator

$$D = Id + A \tag{9.7}$$

is called the *covariant exterior derivative operator* relative to the gauge connection matrix-vector A. The vector components in space-time are

$$D_\mu = I\partial_\mu + A_\mu \tag{9.8}$$

with matrix components $D^a_{\mu b}$ defined in the Lie algebra of the symmetry group.

However, to be consistent with the usual derivative operator and the covariant derivative defined in the geometry of manifolds of Chapter 2 and the exterior derivative defined in Chapter 4, the operator (9.8) must satisfy the formal conditions (which are typical of derivatives):

1) $D \wedge (\alpha\Psi + \beta\Phi) = \alpha D(\Psi) + \beta D(\Phi)$
2) $D \wedge f = df(x)$
3) $D \wedge (f(x)\Psi) = df(x) \wedge \Psi + f(x)D\Psi$
4) $D \wedge (\Psi \wedge \Phi) = (D\Psi) \wedge \Phi + \Psi \wedge (D\Phi)$

where $f(x)$ is a scalar function defined in $\mathscr{M}$.

It is important to observe that A_μ was not really postulated or defined. They are just coefficients of the variation of the Lie algebra basis in terms of the covariant coordinate basis $\{dx^\mu\}$. On the other hand, it was derived from a symmetry group of a Lagrangian, so that it must coincide with the same coefficients of the derivative operator (8.22) defined in Noether's theorem for local gauge transformations, which has also components in the Lie algebra. Therefore, the above connection components $A^a_{\mu b}$ are the same as those introduced by Noether, satisfying Noether's condition (8.21)

$$\sum_a \mathscr{F}_a \frac{\partial\theta^a}{\partial x^\mu} = \sum_a \mathscr{F}_a g A^a{}_{\mu b}\theta^b$$

Consequently the local gauge derivative used in Noether's theorem is the same covariant derivative obtained above from the Lie algebra trivialization of the symmetry group.

Summarizing, we have seen that a principal fiber bundle of a Lie symmetry group G, defined on a space–time $\mathscr{M}$ of some field Ψ, can be trivialized by taking the dual adjoint representation of the Lie algebra of G, $\mathscr{G}^*$, when the total space becomes a product of the base by a typical fiber

$$(\mathscr{M},\ G,\ \pi,\ T\mathscr{B}) \rightarrow (\mathscr{M},\ \pi,\ \mathscr{M} \times \tilde{\mathscr{G}}^*)$$

Here $T\mathscr{B}$ is the total space where the fields are defined as maps $\Psi : \mathscr{M} \rightarrow T\mathscr{B}$.

The trivialization is a direct consequence of the fact that the symmetry group G acts on the field manifold $T\mathscr{B}$ and also on any of its representation space. Such scheme is very general, but in the particular case of the adjoint representation of the Lie algebra of G, the field becomes a vector in the Lie algebra or in its dual $\tilde{\mathscr{G}}^*$.

In the case where the Lie groups are locally defined, the dependence of the parameters on the coordinates implies that the basis of its Lie algebra $\{X_a\}$ defined in (3.7) also depends on the coordinates of $\mathscr{M}$. In such cases the trivialization of the principal bundle by the adjoint representation implies that the field is expressed as

$$\Psi(x^\mu) = \Psi^a(x^\mu) X_a(x^\mu) = \Psi_a(x^\mu)\, X^a(x^\mu)$$

where both the components and the basis vectors depend on the coordinates x^μ. The consequence of this double dependence on the coordinates (of the components and of the basis vectors) is the emergence of an affine connection and of the exterior covariant derivative in local gauge fields defined by

$$D \wedge \Psi = d\Psi_a X^a + \Psi_a \wedge dX^a$$

or

$$D \wedge \Psi = D^a_{\mu b} \Psi_a \, dx^\mu \wedge X^b$$

where $D^a_{\mu b}$ are components of the *exterior covariant derivative* matrix operator

$$D_\mu = I \partial_\mu + A_\mu$$

with entries defined in the adjoint representation of the Lie algebra. Thus, in matrix notation we may write the gauge covariant derivative operator as matrix-vector operator D:

$$D = D_\mu dx^\mu \tag{9.9}$$

We say that the matrix-vector A_μ are the components of the connection associated with the Lie group G.

Since in Noether's theorem $A^b_{\mu a}$ were not really defined we still do not know what the connection is. We only know that it is required to define the conserved quantities for local gauge symmetries. The Yang–Mills theory described in the next section defines such connection.

Chapter 10
Gauge Fields

10.1 Gauge Curvature

The connection associated with local gauge symmetry seen in the last chapter is not complete because its components $A_{\mu a}^{b}$ were not determined. To finish the theory we need to go one step further, by associating with it a curvature and the corresponding field equations.

The concept of curvature in gauge theory is the same introduced by Riemann in 1850. This is not always readily appreciated because there are two different usages for the designation *Riemann tensor*. One is the general definition of the Riemann tensor by a displacement of a vector field around a parallelogram as seen in Chapter 2. The other, more common in the applications to Einstein's equations, is the expression of the Riemann tensor specifically calculated for the metric connection.

The first case is more general because any additional condition imposed on the connection can be made afterward. To avoid confusion some authors refer to the second case as "non-Riemannian geometry" [113], while the designation of Riemannian geometry is reserved for the specific case of a metric connection.

To see that the expression of the Riemann curvature tensor is independent of the choice of connection let us recall from Chapter 2 the expression for the Riemann tensor for the metric connection when we have two linearly independent tangent vector fields. In the particular case of a tangent basis $\{e_\mu\}$ we obtain

$$R(e_\mu, e_\nu) = (\nabla_\mu \nabla_\nu - \nabla_\nu \nabla_\mu) = [\nabla_\mu, \nabla_\nu] \tag{10.1}$$

where we have simplified the notation ∇_{e_μ} to ∇_μ. Notice that this is anti-symmetric in the two indices and consequently it cannot be confused with the Ricci tensor. To avoid notational confusion we denote the *Riemann curvature operator* by

$$\mathscr{R}_{\mu\nu} = [D_\mu, D_\nu] \tag{10.2}$$

This is the same expression for the Riemann curvature (10.1) before the Levi-Civita connection is chosen. In this case, applying the above operator to a basis vector

M.D. Maia, *Geometry of the Fundamental Interactions*,
DOI 10.1007/978-1-4419-8273-5_10,

e_ρ, we obtain from its definition a linear combination of the tangent basis, whose coefficients $R_{\mu\nu\rho}{}^\sigma$ are components of the Riemann tensor in that basis

$$\mathscr{R}_{\mu\nu}\, e_\rho = R(e_\mu, e_\nu)\, e_\rho = R_{\mu\nu\rho}{}^\sigma\, e_\sigma \tag{10.3}$$

These components can be calculated for any given connection in terms of its Christoffel symbols. In the following we define the curvature tensor for a connection defined in a given Lie algebra.

Definition 10.1 (Curvature Two-Form) Consider a vector-matrix one-form gauge connection A defined in space–time, with matrix components defined in the Lie algebra $\mathscr{G}$ of a local gauge symmetry group G. The *curvature two-form* of A is a two-form given by the exterior covariant derivative of A

$$F = D \wedge A = (d + A) \wedge A = d \wedge A + A \wedge A \tag{10.4}$$

The last term has the meaning of the anti-symmetric tensor product of the matrix A by itself, so that it is not necessarily zero [17, 108, 112].

Theorem 10.1

$$F = D \wedge A = F_{\mu\nu} dx^\mu dx^\nu, \quad \textit{where } F_{\mu\nu} = [D_\mu, D_\nu]$$

From the above definition and the properties of the exterior covariant derivative (9.8), we obtain

$$\begin{aligned}
&F = D \wedge A = d \wedge \left(\sum A_\mu dx^\mu\right) + \sum_{\mu<\nu} A_\mu A_\nu dx^\mu \wedge dx^\nu = \\
&\sum_{\mu<\nu} A_{\mu,\nu} dx^\mu \wedge dx^\nu + \sum_{\mu<\nu} A_\mu A_\nu dx^\mu \wedge dx^\nu = \\
&\sum \left[(\partial_\mu A_\nu - \partial_\nu A_\mu) + (A_\mu A_\nu - A_\nu A_\mu)\right] dx^\mu dx^\nu
\end{aligned} \tag{10.5}$$

In these expressions we have used the fact that $F_{\mu\nu}$ is an anti-symmetric tensor, so that the sum $\sum F_{\mu\nu} dx^\mu \wedge dx^\nu$ contains twice the same terms. To avoid the double counting of indices it is sufficient to remove the wedge of the exterior product and unrestrict the sum.[1] This is equivalent to writing two-forms as anti-symmetric

[1] Since any product of indices can be decomposed as half the sum of the symmetrized product and anti-symmetrized product, respectively, $\mu\nu = ((\mu\nu) + [\mu\nu])/2$, the unrestricted sum of a product of a symmetric tensor and an anti-symmetric tensor cancels the symmetric terms automatically.

tensors. Thus, for example, the above expression can be written with an unrestricted Einstein summation convention as

$$F = F_{\mu\nu} dx^{\mu} dx^{\nu}$$

On the other hand, the components of the exterior covariant derivative are $D_\mu = I\partial_\mu + A_\mu$. Defining the *gauge curvature operator*

$$F_{\mu\nu} = [D_\mu, D_\nu]$$

we obtain for an arbitrary function f

$$\begin{aligned}
&F_{\mu\nu} f = [D_\mu, D_\nu] f = D_\mu(D_\nu(f)) - D_\nu(D_\mu(f) = \\
&(\partial_\mu + A_\mu)(\partial_\nu + A_\nu) f - (\partial_\nu + A_\nu)(\partial_\mu + A_\mu) f - \\
&\partial_\mu \partial_\nu f + \partial_\mu A_\nu f + A_\nu \partial_\mu f + A_\mu \partial_\mu f - A_\mu A_\nu f - \\
&(\partial_\nu \partial_\mu f + \partial_\nu A_\mu f + A_\mu \partial_\nu f + A_\nu \partial_\nu f - A_\nu A_\mu f) = \\
&\big(\partial_\mu A_\nu - \partial_\nu A_\mu) f + [A_\mu, A_\nu]\big) f
\end{aligned}$$

Removing f and comparing the last row with the last row of (10.5), we conclude that

$$F = D \wedge A = F_{\mu\nu} dx^{\mu} dx^{\nu}, \qquad F_{\mu\nu} = [D_\mu, D_\nu] \tag{10.6}$$

Therefore the *concept of gauge curvature is the same as the Riemann curvature operator, differing only by the choice of connection.*

Theorem 10.2

$$D \wedge F = 0$$

Taking the exterior covariant derivative of (10.1) and using the properties of D_μ, we obtain a three-form

$$\begin{aligned}
&D \wedge F = D \wedge (D \wedge A) = \sum_{\mu<\nu<\rho} (D_\rho F_{\mu\nu}) dx^{\rho} \wedge dx^{\mu} \wedge dx^{\nu} = \\
&\sum_{\mu<\nu<\rho} D_\rho [D_\mu, D_\nu] dx^{\rho} \wedge dx^{\mu} \wedge dx^{\nu}
\end{aligned}$$

Remembering the anti-symmetry properties of the factors we may write this result as

$$D \wedge F = \sum [D_\rho, [D_\mu, D_\nu]] dx^{\rho} dx^{\mu} dx^{\nu}$$

However, the three-dimensional anti-symmetrization symbol applies to any fixed sequence of indices (and not necessarily to just 1,2,3). That is, $\varepsilon^{\mu\nu\rho}_{\alpha\beta\gamma}$ is 1 for even permutations of (μ, ν, ρ) with respect to (α, β, γ); -1 for odd permutations; and 0 in any other case. Thus, and we may write the above expression as

$$[D_\rho, [D_\mu, D_\nu]] = \varepsilon^{\alpha\beta\gamma}_{\mu\nu\rho}[D_\alpha, [D_\beta, D_\gamma]]$$

Therefore,

$$D \wedge F = \sum \varepsilon^{\rho\mu\nu}_{\alpha\beta\gamma}[D_\rho, [D_\mu, D_\nu]]dx^\alpha dx^\beta dx^\gamma$$

Since D_α is an operator defined in a Lie algebra, the Jacobi identity

$$[A, [B, C]] + [C, [A, B]] + [B, [C, A]] = 0$$

applies for any cyclic choice of the triad of indices (α, β, γ). Therefore we may write

$$\sum \varepsilon^{\rho\mu\nu}_{\alpha\beta\gamma}[D_\rho, [D_\mu, D_\nu]] = [D_\alpha, [D_\beta, D_\gamma]] + [D_\gamma, [D_\alpha, D_\beta]] + [D_\beta, [D_\gamma, D_\alpha]] = 0$$

Consequently,

$$D \wedge F = 0 \tag{10.7}$$

From (10.4) the above expression can also be written as $D \wedge D \wedge A = 0$, which is sometimes written as $D^2 A = 0$.

This is as far as we may go without specifying the connection A_μ. Just as a reminder, it appeared for the first time in Noether's theorem for local gauge fields and later on, we understood that the same quantity acts as a connection which is required to make the Lagrangian invariant. Equation (10.7) by itself is not sufficient to determine A_μ because it is in fact just an identity.

In order to make further progress, we need to specify an equation for the connection and the Lie symmetry that defines it. In contrast with general relativity where the connection is postulated, here the connection is a dynamical field in itself, called the *gauge field*. The development of this concept started in 1954 with Yang and Mills, and as it turned out, it is responsible for the description of three of the four fundamental interactions, named after the three gauge symmetry groups, $U(1)$, $SU(2)$, and $SU(3)$. In the following we will examine each of these three gauge symmetries, starting with the electromagnetic theory.

10.2 The $U(1)$ Gauge Field

The $U(1)$ gauge field is the electromagnetic theory, described as connection theory of the local $U(1)$ group. As such, it serves as a paradigm for all other gauge theories.

It is relevant to remind that the electromagnetic theory is the result of the intense experimental research as described in Chapter 7. We have also seen in the same chapter that the infinitesimal gauge transformations of the electromagnetic potential can be written as an infinitesimal rotation of the local $SO(2)$ group. This group is isomorphic to the unitary group $U(1)$, giving the infinitesimal transformations of the electromagnetic four-potential as

$$A'_{\mu} = A_{\mu} + i\theta_{,\mu} \tag{10.8}$$

On the other hand, we have also seen in the previous chapter how a generic gauge connection associated with a Lie-type symmetry group G can be derived by the trivialization obtained by the adjoint representation of the Lie algebra. This gave us a general recipe for deriving covariant derivatives of fields for any field theory.

Therefore, in principle we may derive the electromagnetic potential as a connection, by constructing the dual adjoint representation of the Lie algebra of the electromagnetic gauge group $U(1)$. As it happens, the Lie algebra $\mathscr{G}_{U(1)}$ has only one basis element X which can be expressed as

$$X = a^{\mu}\frac{\partial}{\partial x^{\mu}}, \quad \text{with dual } \; X^* = a_{\mu}dx^{\mu} \tag{10.9}$$

Therefore, the Lie algebra of that group, $\mathscr{G}_{U(1)}$, is Abelian because the Lie product vanishes: $[X, X] = 0$. This means that all of its structure constants also vanish, and we cannot derive the electromagnetic potential directly from it.

However, from (10.9) we see also that the only surviving Lie algebra operator X can be written as a linear combination of the (coordinate) basis, either of the tangent or of the cotangent bundle of Minkowski's space–time $T_p\mathscr{M}$. Therefore, expression (10.9) indicates that the principal bundle of $U(1)$ can also be trivialized by a base morphism of a fiber bundle in which the total space is the Cartesian product between the base space $\mathscr{M}$ and the dual of Minkowski's tangent space–time.

$$(\mathscr{M}_4,\ U(1),\ \pi,\ T\mathscr{B}) \rightarrow (\mathscr{M}_4,\ \pi,\ \mathscr{M}_4 \times \mathscr{M}^*)$$

Actually this is an example of equivalent trivializations, producing equivalent representations of the group $U(1)$. O one is the adjoint representation and the other is a 1×1 irreducible unitary representation of a subgroup of the Lorentz group.

In Chapter 3 we have seen that we may define representations of a Lie algebra with different dimensions, but they may be decomposed into smaller representations up to a certain size, when they are called irreducible matrix representations. For example, a representation of a Lie algebra may be formed by diagonal block matrices of different sizes, $[A]$, $[B]$, etc., so that the representation may decompose as a direct product of smaller representations

$$\begin{pmatrix} [A] & & & \\ & [B] & & \\ & & [C] & \\ & & & [D] \end{pmatrix} = [A] \otimes [B] \otimes [C] \otimes [D]$$

In particular when the Casimir operator of smallest order of a semi-simple group is proportional to the identity matrix, then the above decomposition is completely reducible, when the diagonal blocks are also diagonal. (This is one of the lemmas of Schur; see, e.g., [63].)

In the case of a semi-simple group we obtain an Abelian 1×1 representation, although it is not constructed with the structure constants. Thus, the connection one-form associated with this irreducible representation is naturally written in coordinate basis as (10.9). A is a 1×1 matrix one-form where its components A_α are real functions

$$A = A_\mu dx^\mu$$

The covariant derivative associated with this connection is also written as $D = D_\mu dx^\mu$ where $D_\mu = \partial_\mu + A_\mu$. From the fact that A is a 1×1 matrix it follows that $[A_\alpha, A_\beta] = 0$. Consequently

$$A \wedge A = \sum_{\alpha<\beta} A_\alpha A_\beta dx^\alpha \wedge dx^\beta = [A_\alpha, A_\beta] dx^\alpha dx^\beta = 0$$

Hence,

$$D \wedge A = d \wedge A + [A, A] = d \wedge A \tag{10.10}$$

and the curvature of this connection is

$$F = D \wedge A = d \wedge A = (\partial_\alpha A_\beta - \partial_\beta A_\alpha) dx^\alpha dx^\beta \tag{10.11}$$

so that

$$F_{\alpha\beta} = \partial_\alpha A_\beta - \partial_\beta A_\alpha \tag{10.12}$$

which are the components of Maxwell's tensor, provided they satisfy Maxwell's equations. This can be seen from the following:

Taking the covariant derivative of F according to (10.11), we obtain a three-form (here it may be easier to follow the arguments using the explicit exterior product)

$$D \wedge F = \sum_{\alpha<\beta<\gamma} F_{\alpha\beta,\gamma}\, dx^\gamma \wedge dx^\alpha \wedge dx^\beta$$

However, in four dimensions a three-form is isomorphic to a one-form, as given by the relation

$$dx^\gamma \wedge dx^\alpha \wedge dx^\beta = \varepsilon^{\gamma\alpha\beta\delta} dx^\delta \tag{10.13}$$

so that we may write the previous expression as

$$D \wedge F = \sum \varepsilon^{\gamma\alpha\beta\delta} F_{\gamma\alpha,\beta} dx^\delta$$

Now, applying the homogeneous Maxwell's equations (7.15) $\varepsilon^{\gamma\alpha\beta\delta} F^{\beta\delta}{}_{,\beta} = 0$, it follows that

$$D \wedge F = 0$$

This is the same result expressed as (10.7) which holds independently of the dynamical principle and is equivalent to the Jacobi identity. It corresponds to the two homogeneous Maxwell's equations (Biot–Savart and Faraday).

On the other hand, the dual of the Maxwell tensor F^* is defined by the components

$$F^*_{\alpha\beta} = \varepsilon_{\alpha\beta\gamma\delta} F^{\gamma\delta} \tag{10.14}$$

so that F^* is a dual curvature two-form

$$F^* = \sum F^*_{\gamma\delta}\, dx^\gamma \wedge dx^\delta = \sum \varepsilon_{\alpha\beta\gamma\delta} F^{\gamma\delta} dx^\alpha \wedge dx^\beta$$

Using (10.10), the covariant derivative of this dual is again a three-form

$$D \wedge F^* = \sum \varepsilon_{\alpha\beta\gamma\delta} F^{\gamma\delta}{}_{,\mu} dx^\mu \wedge dx^\alpha \wedge dx^\beta$$

which in four dimensions (and only in four dimensions) is also isomorphic to a one-form

$$D \wedge F^* = \sum \varepsilon_{\alpha\beta\gamma\delta} F^{\gamma\delta}{}_{,\mu} \varepsilon^{\mu\alpha\beta\nu} dx^\nu$$

or using the properties of the four-dimensional anti-symmetrization symbol: $\varepsilon^{\alpha\beta\gamma\delta}$ is equal to 1 for even combinations of 1234; to -1 for odd combinations; and 0 in any other cases, it follows that $\varepsilon_{\alpha\beta\gamma\delta}\varepsilon^{\mu\alpha\beta\nu} = \delta^\mu_{[\gamma}\delta^\nu_{\delta]}$.

Therefore,

$$D \wedge F^* = \sum F^{\mu\nu}{}_{,\nu} dx^\nu$$

Now, applying the non-homogeneous Maxwell's equations from (7.14) we obtain

$$D \wedge F^* = 4\pi J$$

where J denotes the electric current four-vector in Minkowski's space–time.

We conclude that Maxwell's electromagnetic theory is the gauge field theory for the $U(1)$ gauge group, whose connection is the electromagnetic four-vector defined in Minkowski's space–time, satisfying Maxwell's equations

$$\begin{cases} D \wedge F = 0 \\ D \wedge F^* = 4\pi J \end{cases} \tag{10.15}$$

The electric and magnetic fields are the components of the curvature two-form F.

The Lagrangian of the electromagnetic field in terms of the Maxwell curvature operator is given by (7.16):

$$\mathscr{L} = \frac{1}{4} F^{\mu\nu} F_{\mu\nu} \tag{10.16}$$

where $F_{\mu\nu}$ is given by (10.12). As we see, there are only kinetic terms in this Lagrangian. Comparing with the Klein–Gordon equation, it does not have a mass term. Since this Lagrangian is invariant under the electromagnetic gauge transformations, we cannot recover a mass by such transformations. This means that the connection A_μ is a massless field. This will be also a feature of the next gauge fields.

10.3 The $SU(2)$ Gauge Field

In 1954 Yang and Mills proposed a generalization of the local $U(1)$ gauge theory, where the gauge group would be replaced by the local $SU(2)$ group. The motivation in doing so was an attempt to explain the isospin. Differently from the global isospin, the proposed distinction between protons and neutrons should be local, different at each point. This would be similar to the local gauge symmetry of the electromagnetic field, but with the local $SU(2)$ gauge symmetry.

Consider a two-component spinor field defined in the two-dimensional complex spinor space, on which the $SU(2)$ Lie group acts as a local transformation group $\mathscr{S}_2$,

$$\Psi = \begin{pmatrix} \Psi_1 \\ \Psi_2 \end{pmatrix}$$

The transformation of the local $U(1)$ group can be summarized as

$$\Psi' = u\Psi = F(\Psi, \theta) \tag{10.17}$$

The $SU(2)$ group is composed of 2×2 matrix operators with parameters a_{ab}

$$u = \begin{pmatrix} a_{11} & a_{12} \\ a_{21} & a_{22} \end{pmatrix}$$

to which we add the unitary and unimodular conditions, respectively, described by $uu^\dagger = 1$ and $\det u = 1$. They give the following equations for the parameters:

$$\begin{aligned} |a_{11}|^2 + |a_{12}|^2 &= 1 \\ |a_{21}|^2 + |a_{22}|^2 &= 1 \\ a_{11}a_{21}^* + a_{12}a_{22}^* &= 0 \\ a_{11}a_{22} - a_{12}a_{21} &= 0 \end{aligned}$$

From these equations we conclude that there are only three independent parameters. This Lie group is isomorphic to $SO(3)$, also with three parameters. Such 1:1 correspondence with rotations resembles the quantum mechanical spin, with the difference that here this $SO(3)$ results from a field transformation, instead of the coordinate transformation.

From the general theory of connections described in the previous chapter, the adjoint representation of the local Lie algebra $\mathcal{G}_{SU(2)}$ trivializes the principal fiber bundle of $SU(2)$ to

$$(\mathcal{M}_4,\ \pi,\ \mathcal{M}_4 \times \tilde{\mathcal{G}}^*_{4SU(2)})$$

As it happened in the electromagnetic case, we may use another more convenient group to realize the trivialization. Here, for example, we may use the isomorphism between $SU(2)$ and $SO(3)$ and the isomorphism between the $SO(3)$ and the *group of automorphisms of the quaternion algebra* $\mathbb{C}_2$ described in Chapter 7, which generates as the group of transformations of two-component spinors. Therefore, by this double isomorphism we find that the local $SU(2)$ spinors are the same spinors associated with the local two-component spinors derived from the local quaternion matrix representation.

We have a similar situation with the previous example: the action of the group on the $SU(2)$ field is the same as the quaternion spinor representation, which in turn is the same as the action of the group of automorphism of the quaternion algebra in the matrix representation. As such they are vectors of the matrix representation of the $\mathbb{C}_2$ algebra written in the same basis given by the Pauli matrices. It must be emphasized, however, that these Pauli matrices do not refer to the orbital spin of non-relativistic quantum mechanics, but to a *relativistic internal quantum number*, which is the intended generalization of the global isospin to the *local isospin*.

In order to avoid confusion with the usual Pauli matrices of the orbital spin, we use a different notation as follows:

$$\tau_0 = \begin{pmatrix} 1 & 0 \\ 0 & 1 \end{pmatrix},\ \tau_1 = \begin{pmatrix} 0 & 1 \\ 1 & 0 \end{pmatrix},\ \tau_2 = \begin{pmatrix} 0 & -i \\ i & 0 \end{pmatrix},\ \tau_3 = \begin{pmatrix} 1 & 0 \\ 0 & -1 \end{pmatrix} \qquad (10.18)$$

satisfying the same multiplication table for the $SO(3)$ Pauli matrices

$$\tau_i\tau_j + \tau_j\tau_i = 2\delta_{ij}\tau_0, \quad i, j = 1, \ldots, 2$$
$$\tau_1\tau_2 = \tau_3$$

From the isomorphism between the Lie algebra $\mathcal{G}_{SU(2)}$ and $SO(3)$ and the isomorphism between the later and a subgroup of the automorphisms of quaternions, the adjoint representation of $\mathcal{G}_{SU(2)}$ can be expressed in terms of the above matrices. Including the identity matrix τ_0, we may express any operator of the adjoint representation of the $\mathcal{G}_{SU(2)}$ by τ_i:

$$\Psi = \Psi^0\tau_0 + \sum_1^3 \Psi^i\tau_i = \sum_0^3 \Psi^\mu\tau_\mu = \sum_0^3 \Psi_\mu\tau^\mu$$

where indices are risen and lowered by the Minkowski metric.

Likewise, the gauge potential A and the gauge covariant derivative are represented as operators in the same Lie algebra, written in terms of the isospin matrices basis $\{\tau_\mu\}$. Then the proposition of Yang and Mills was to write the field equations for A similar to Maxwell's, now called the *Yang–Mills equations*. The solutions of these equations give the connection components A_μ, from which we obtain the $SU(2)$ connection as a quaternion field

$$A = \sum_0^3 A^\mu\tau_\mu = \sum_0^3 A_\mu\tau^\mu$$

Unlike Maxwell's potential, these are 2×2 matrices which do not commute with each other.

Therefore, the Lagrangian of this generalization of the electromagnetic field is similar to (10.16). Since now the curvature tensor $F_{\mu\nu}$ is a matrix and the Lagrangian by definition a scalar functional of the field, and its derivatives, Yang and Mills proposed that we should take the trace of the resulting matrix:

$$\mathcal{L}_{SU(2)} = \frac{1}{4} tr\, F_{\mu\nu}F^{\mu\nu} = \frac{1}{4} tr\, \eta^{\mu\rho}\eta^{\nu\sigma}F_{\mu\rho}F_{\nu\sigma} \tag{10.19}$$

where $F_{\mu\nu}$ is

$$F_{\mu\nu} = [D_\mu, D_\nu] = \partial_\mu A_\nu - \partial_\nu A_\mu + [A_\mu, A_\nu]$$

The Euler–Lagrange equations are expressed in the usual functional way as

$$\frac{\partial \mathcal{L}_{SU2}}{\partial A_\rho} = \frac{\partial}{\partial x^\sigma}\frac{\partial \mathcal{L}_{SU(2)}}{\partial A_{\rho,\sigma}}$$

When calculating the derivatives with respect to a non-commutative variable we need to be careful with the order of the factors. Considering the order and the trace we obtain the Euler–Lagrange equations

$$F^{\mu\nu}{}_{,\nu} = [A_\nu, F^{\mu\nu}] \tag{10.20}$$

Comparing with the non-homogeneous Maxwell equations (7.17), we see that the right-hand side plays the role of the current J^μ. This term does not appear in the $U(1)$ theory because the potential A_μ commutes with everything. The term $[A_\nu, F^{\mu\nu}]$ in (10.20) is typical of the non-linearity of the Yang–Mills for non-Abelian gauge fields and it is called the *Yang–Mills current*. The result was at the time of its proposal something entirely new, within the context of the weak nuclear interaction and its unification with the electromagnetic interaction, known as the *electroweak unification*.

However, it is also possible to have a solution for the $SU(2)$ Yang–Mills field without any current, called the *instanton*.

Example 10.1 (Instantons) Consider an $SU(2)$ gauge field with an additional condition

$$F^* = \pm F$$

called the self-dual and anti-self-dual conditions, respectively. In any of these cases the equations become similar to the vacuum Maxwell's equations

$$D \wedge F = 0 \text{ and } D \wedge F^* = 0 \tag{10.21}$$

In spite of this simplification the equations remain non-linear because the matrices A_μ do not commute. For the relation between instantons and mathematics and applications, see [114–117].

A particular solution of the $SU(2)$ Yang–Mills equations using the quaternion basis (10.18) was presented by M. F. Atiyah. The idea is to find an appropriate quaternion function whose differential provides the $SU(2)$ connection satisfying the (10.21) conditions [118].

To see how this works, consider a point in Minkowski's space–time with coordinates x^μ. It corresponds to a quaternion variable, written in the Pauli basis as $X = \sum_0^3 x^\mu \tau_\mu$. In particular Atiyah considered two quaternion functions

$$f(X) = \frac{\bar{X}}{1+|X|^2} = \frac{x^0\tau_0 - \sum_1^3 x^i \tau_i}{1 + x^{0^2} + \sum_1^3 x^{i^2}} \tag{10.22}$$

and its quaternion conjugate

$$\bar{f}(X) = \frac{X}{1+|X|^2} = \frac{x^0\tau_0 + \sum_1^3 x^i \tau_i}{1 + x^{0^2} + \sum_1^3 x^{i^2}} \tag{10.23}$$

such that the $SU(2)$-gauge potential is the differential $df(X)$ of these functions. However, since this potential corresponds to a vector in space–time, we need to extract the vector components of the quaternion $df(X)$. This is obtained by simply taking the imaginary components of the quaternion differential of the functions $f(X)$:

$$A(X) = \frac{1}{2}(df(X) - d\bar{f}(X))$$

However, quaternion analysis is non-commutative. That is, left and right differentials are not equal and can only be understood within the context of non-commutative field theory, of which quaternion analysis is a particular example [119–121]. Therefore, when expressing in terms of quaternions we may also have left and right $SU(2)$ gauge fields. The left derivative of a quaternion function can be defined by the traditional limit, using the mentioned division algebra property:

$$f'(X) = \lim_{\Delta X \to 0} \left[f(X + \Delta X) - f(X)\right] \Delta X^{-1}$$

and similarly for the right derivative is

$$'f = (X) \lim_{\Delta X \to 0} \Delta X^{-1} \left[f(X + \Delta X) - f(X)\right]$$

These are generally not equal (see, e.g., [122, 123] and references therein).

In view of this, the left differential of a quaternion function can be defined as

$$df(x) = f'(x)dx$$

Using the Pauli basis this can be written as

$$\begin{aligned} df = {} & (U^0\tau_0 + U^a\tau_a)(dx^0\tau_0 + dx^b\tau_b) = \\ = {} & (U^0 dx^0 + U^a dx^a)\tau_0 + (U^0 dx^a + U^a dx^0 + \varepsilon_{abc} U^b dx^c)\tau_a \end{aligned}$$

Thus the left vector-matrix connection is given by the vector part of this left differential:

$$A(X) = \frac{1}{2}(df(X) - d\bar{f}(X)) = \sum A_\mu(X) dx^\mu = A_0 dx^0 + A_i dx^i$$

where the components A_μ are expressed in terms of the 2×2 matrices of the $SU(2)$ Lie algebra as

$$A_0 = \mp \sum \frac{x^i \tau_i}{1 + |X|^2} \quad \text{and} \quad A_j = \sum \frac{x^0 \tau_j}{1 + |X|^2} \mp \frac{\varepsilon_{ijk} x^i \tau_k}{1 + |X|^2}$$

With these components we construct the $SU(2)$ covariant derivative of the operators $D_\mu = I\partial_\mu + A_\mu$ and the corresponding curvature two-form $F = F_{\mu\nu}dx^\mu dx^\nu$, with $F_{\mu\nu} = [D_\mu, D_\nu]$. After some algebra, denoting $dX = dx^\alpha\tau_\alpha$, we obtain

$$F = \mp\frac{dX \wedge d\bar{X}}{(1+|X|^2)^2}$$

where (considering the restricted sums on the wedge products)

$$dX \wedge d\bar{X} = -(dx^0 \wedge dx^i\tau_i + dx^i \wedge dx^j\varepsilon_{ijk}e^k)$$

Therefore,

$$F = \pm\frac{dx^0 \wedge dx^i\tau_i}{(1+|X|^2)^2} \pm \frac{\varepsilon_{ijk}dx^i \wedge dx^k\tau_k}{(1+|X|^2)^2}$$

so that the components $F_{\mu\nu}$ are

$$F_{0i} = \frac{\pm dx^0 \wedge dx^i\tau_i}{(1+|X|^2)^2} \text{ and } F_{ij} = \frac{\pm\varepsilon_{ijk}\tau_k}{(1+|X|^2)^2}$$

We find that indeed the proposed functions represent solutions of the $SU(2)$ Yang–Mills equations such that $F = \pm F^*$ [118].

The emergence of the unification scheme of Weinberg and Salam, called the *electroweak unification*, invariant under the $U(1) \times SU(2)$ gauge group, helped to understand the meaning of the Yang–Mills proposition [24, 25]. The original derivation of the electroweak unification was based on phenomenological arguments, but it is compatible with the general theory of connections. Essentially, consider a relativistic electron interacting with a $U(1)$ field in the Minkowski space–time and interacting with a Yang–Mills $SU(2)$ gauge field.

The Lagrangian of this system is composed of the Dirac Lagrangian

$$\mathscr{L} = \mathscr{L}_{SU(2)}(A_\mu, A_{\mu,\rho}) + \mathscr{L}_{\text{Dirac}}(\Psi, \Psi_{,\mu})$$

which is invariant under the $U(1)$ local gauge group and a $SU(2)$ invariant Yang–Mills Lagrangian

$$\mathscr{L}_{SU(2)} = \frac{1}{4}tr F_{\mu\nu}F^{\mu\nu}$$

The Dirac Lagrangian is

$$\mathscr{L}_{\text{Dirac}} = i\bar{\Psi}\gamma^\mu\partial_\mu\Psi - m\bar{\Psi}\Psi$$

Apparently there is no interaction term. However, like in the Nielsen–Olesen example an explicit interaction term may emerge when we rewrite the total Lagrangian with the appropriate covariant derivative, derived from the combined symmetry $U(1) \times SU(2)$.

Denoting the gauge potential of the combined symmetry by $A_{,\mu}$, and the corresponding covariant derivative by $D_\mu = i\partial_\mu + A_\mu$, we may rewrite the total Lagrangian in the invariant form as

$$\mathscr{L}_{\mathrm{EW}} = \frac{1}{4} tr\, F_{\mu\nu} F^{\mu\nu} + i\bar{\Psi}\gamma^\mu D_\mu \Psi - m\bar{\Psi}\Psi + HC$$

where $F_{\mu\nu} = [D_\mu, D_\nu]$.

The Euler–Lagrange equations with respect to Ψ give the Dirac equation

$$(i\gamma^\mu \partial_\mu - m)\Psi + i\gamma_\mu A_\mu \Psi = 0$$

where we notice the emergence of the interaction term involving $A_\mu \Psi$. On the other hand, the Euler–Lagrange equations with respect to A_μ give the Yang–Mills equation for the $SU(2)$ field with a Dirac current

$$tr\, F^{*\rho\sigma}{}_{,\sigma} = 2itr\, \bar{\Psi}\gamma^\rho \Psi$$

However, since we have non-Abelian local symmetry we would expect to see also the emergence of the Yang–Mills current.

The Yang–Mills current is hidden because the trace was unduly maintained in the Yang–Mills equation. This can be corrected by taking the exterior product of the above equation by dx^ρ and adding to both sides of the result the trace of the Yang–Mills current $tr\,(dx^\rho \wedge [A_\sigma, F^{*\rho\sigma}])$, obtaining

$$tr\, dx^\rho \wedge (F^{*\rho\sigma}{}_{,\sigma} + [A_\sigma, F^{*\rho\sigma}]) = 2itr\, dx^\rho \bar{\Psi}\gamma^\rho \Psi + tr\, dx^\rho \wedge [A_\sigma, F^{*\rho\sigma}]$$

or, equivalently,

$$tr\, dx^\rho \wedge D_\sigma F^{*\rho\sigma} = 2itr\, dx^\rho \wedge \left(\bar{\Psi}\gamma^\rho \Psi + \frac{1}{2i}[A_\sigma, F^{*\rho\sigma}] \right)$$

Now, removing the trace of this equation we obtain a more general equation

$$dx^\rho \wedge D_\sigma F^{*\rho\sigma} = 2idx^\rho \wedge \left(\bar{\Psi}\gamma^\rho \Psi + \frac{1}{2i}[A_\sigma, F^{*\rho\sigma}] \right)$$

so that we obtain the equation for A_μ including the Yang–Mills current. The bracket in the right-hand side gives the total (electroweak) current

$$4\pi J^*_{\mathrm{EW}} \stackrel{\mathrm{def}}{=} 2idx^\rho \wedge \left(\bar{\Psi}\gamma^\rho \Psi + \frac{1}{2i}[A_\sigma, F^{*\rho\sigma}] \right)$$

so that the $U(1) \times SU(2)$ Yang–Mills equation simplifies to

$$D \wedge F^* = 4\pi J^*_{\text{EW}} \tag{10.24}$$

The current J^*_{EW} is conserved in the same sense of Noether's theorem, with zero divergence:

$$D_\alpha D_\beta F^{*\alpha\beta} = 4\pi D_\alpha J^{*\alpha}_{\text{EW}} = 0$$

As in the other cases, the homogeneous equations correspond to the Bianchi identity

$$D \wedge F = 0 \tag{10.25}$$

The Yang–Mills equation (10.24) can now be solved for A and the corresponding field strength can be experimentally verified. The unification holds for energies of the order of 100 GeV.

The Weinberg–Salam theory predicted the existence of the $W^\pm$ and the Z^0 bosons, which were found later on in collisions involving protons and antiproton (see, e.g., [18, 19]).

10.4 The $SU(3)$ Gauge Field

The $SU(3)$ Yang–Mills theory is a result of the development of the quark model for strong interactions. Quarks appeared after an extensive period of modeling the strong nuclear interaction.

In 1964 Gell-Mann and Ne'eman proposed that the interaction within the nucleon should have a symmetry similar to that of the $SU(3)$ group, with an added property called color (often denoted $SU(3)_c$). Therefore, a theory of strong nuclear interactions could in principle be described as a gauge theory of the $SU(3)_c$ local symmetry.

The $SU(3)$ gauge theory is a theory in the making with some open problems, like for example the confinement of quarks and the determination of the minimum mass of the glue-balls. The consistency of the entire gauge theory depends on the solution these problems [123].

Just like the $U(1)$ and the $SU(2)$ cases, at least in principle we may derive the equations for the $SU(3)$ group gauge theory. It is a Lie group with a Lie algebra $\mathscr{G}_{SU(3)}$ generated by 3×3 complex unitary matrices with determinant 1 and dimension 8.

Following the general scheme the connection for such gauge theory can be obtained from the trivialization of the principal bundle of $SU(3)$ by its dual adjoint representation

$$(\mathscr{M}_4,\ SU(3),\ \pi,\ T\mathscr{B}) \rightarrow (\mathscr{M}_4,\ \pi,\ \mathscr{M}_4 \times \tilde{\mathscr{G}}^*_{SU(3)})$$

In such trivialization, the algebra acts on the field and at the same time on the space of the Lie algebra. With this double role played by the adjoint representation the gauge field can be written as a vector-matrix of the same space of the algebra. Thus, denoting by $\{X_a\}$ the basis of the Lie algebra $\tilde{\mathscr{G}}_{SU(3)}$ and its dual by $\{X^a\}$, we may write the $SU(3)$ invariant field as

$$\Psi = \Psi^a X_a = \Psi_a X^a, \quad a = 1, \ldots, 8$$

and the $SU(3)$ gauge exterior covariant derivative of Ψ can be written as

$$D \wedge \Psi = D_\mu \Psi \wedge dx^\mu$$

where D_μ has components $D^b_{\mu a} = \delta^b_a \partial_\mu - g A^b_{\mu a} x$ and where $A^b_{\mu a}$ are the components of the $SU(3)$ connection A_μ defined in space–time.

In the two previously seen gauge fields, the $U(1)$ electromagnetic field and the $SU(2)$ weak interaction field, we have expressed the solutions in terms of division algebras, respectively the complex and the quaternion algebras. The division property was important to define the mathematical analysis and in the determination of solutions. Is this a mere coincidence, or would the fundamental gauge interactions have to do with the structure of division algebras?

In the light of this question it seems logical to revise the next division algebra, the octonion algebra $\mathbb{O}$ (which is not a Clifford algebra), for a possible description of the $SU(3)$ gauge field. This has been proposed in the past by several authors; the result is less clear than in the quaternion case for the $SU(2)$ gauge field [124–126]. The interesting aspect to be considered is that the group of automorphisms of the octonion algebra is a subgroup of the *exceptional Lie algebra* G_2, the smallest among the known exceptional Lie algebras.

The *octonion algebra* is the largest of the normed division algebras, with seven generators plus a unit element, satisfying the multiplication table (where $A, B = 1, \ldots, 7$)

$$\begin{aligned}
&e_1 e_2 = e_3, \quad e_5 e_1 = e_6, \quad e_6 e_2 = e_5 \\
&e_4 e_7 = e_1, \quad e_6 e_7 = e_3, \quad e_5 e_7 = e_2 \\
&e_A e_B + e_B, e_A = -2\delta_{AB} \\
&e_A e_0 = e_0, \ e_A
\end{aligned}$$

This implies also that (as it occurs with the Lie algebras) the octonion algebra is non-associative:

$$[e_A, [e_B, e_C]] \neq [[e_A, e_B], e_C]$$

Instead, it satisfies the Jacobi-like relation

$$[e_A, [e_B, e_C]] + [e_C, [e_A, e_B]] + [e_B, [e_C, e_A]] = 0$$

An octonion is written in the above basis as

$$X = X^0 e_0 + X^i e_i, \quad i = 1, \ldots, 7$$

with conjugate (similar to the quaternion algebra) $\bar{X} = X^0 e_0 - X^i e_i$ and the norm of an octonion is

$$||X||^2 = X\bar{X} = X_0{}^2 + \sum_0^7 X^{A^2}$$

The analytical conditions for the existence of analytical functions of octonions (in the sense of the Cauchy–Riemann conditions) are even more restrictive than those for quaternions. In view of this, like in the case of quaternions we also consider separate left and right derivatives of octonion functions.

The group of automorphisms of the octonion algebra is a 14-parameter exceptional Lie group denoted by G_2, which is the smallest among the exceptional Lie groups [61]. Therefore, the group of automorphisms of the octonions is not isomorphic to the gauge group $SU(3)$. Nonetheless, it is possible to fix one of the octonion basis elements to obtain seven possible subalgebras, each of which has a subgroup of automorphisms isomorphic to $SU(3)$ [124]. For example, by excluding (or fixing) e_7 in the above multiplication table, we obtain a subalgebra of the octonions with seven generators. Denoting by $f_1, \ldots, f_7$ these generators Gell-Mann showed that they satisfy a Lie-like product

$$[f_a, f_b] = \bar{f}_{abc} f_c$$

where $\bar{f}_{abc}$ are called the *Gell-Mann structure constants*, as they keep a correspondence with the $SU(3)$ Lie algebra. Thus, we choose one among the seven possible Gell-Mann sub-algebras of the octonion algebra to represent the $SU(3)$ symmetry.

In any of these subalgebras the $SU(3)$ field is written in the Gell-Mann basis and its dual f^i, plus the identity element f^0 is written as

$$\Psi = \sum_0^7 \Psi^a f_a$$

The $SU(3)$ gauge covariant derivative of this field, written in the same basis, is

$$D \wedge \Psi = (Id + A) \wedge \left(\sum_0^7 \Psi^a f_a \right)$$

satisfying the same general rules for exterior covariant derivatives. We may express this derivative in the dual coordinate basis $\{dx^\mu\}$ as

$$D \wedge \Psi = \sum (\delta_b^a \partial_{,\mu} + A^a_{\mu b}) \Psi_a \, dx^\mu \wedge f^b$$

where $A^b_{\mu a}$ are the matrix components of the $SU(3)$ gauge potential in the Gell-Mann basis. The $SU(3)$ field strengths $F_{\mu\nu} = [D_\mu, D_\nu]$ and the dual $F^{*\mu\nu}$ are expressed in the same Gell-Mann basis.

Therefore, in principle we may write the Yang–Mills equations and hopefully solve them in A_μ written in the octonion subalgebra. This may not be so simple because of the non-linearity, because of the large algebra involved, and because we still need to decide what to do with the remaining six subalgebras.

Following the example of the electroweak theory, it is possible that the solution of these difficulties can be obtained by combining the $SU(3)$ symmetry with the other gauge theories, and perhaps even with gravitation [129].

Exercise ($SU(3)$ Instantons) Following the example of the $SU(2)$ field, verify the existence of self-dual and anti-self-dual $SU(3)$ gauge fields.

Suggestion: following a procedure similar to the case of quaternions devised by Atiyah, look for an octonion function like $F(X) = \frac{\bar{X}}{1+|X|^2}$, written in the Gell-Mann basis. Then check against the $SU(3)$ Yang–Mills vacuum equations $\mathrm{D} \wedge \mathrm{F} = 0$ and $\mathrm{D} \wedge \mathrm{F}^* = 0$.

Chapter 11
Gravitation

11.1 The Riemann Curvature

The gravitational interaction is at the same time the simplest and the most complicated interaction as compared with the three other fundamental interactions. This apparent paradox exists because gravitation does not seem to fit in the same scheme of the gauge interactions as described in the preceding chapters. Contrarily to the gauge theories, Einstein's gravitation has not been quantized, either from the canonical, or from the perturbative points of view [41, 127]. On the side of classical physics the theory can describe the gravitational field of only about 4% of the known universe. The remaining 96% produces a gravitational effect that is not included in Einstein's theory of gravitation.

Einstein's gravitation provided the *geometrical paradigm* from which we have modeled the other interactions. Indeed, as we have seen in Chapters 2, 5, and 10, the basic structure present in all fundamental interactions is the *Riemann tensor* written in coordinate independent form as a linear operator defined by two linearly independent vector fields U, V as

$$\mathscr{R}(U,V)W = [\nabla_U, \nabla_V]W \tag{11.1}$$

where ∇ is the covariant derivative of the Levi-Civita connection.

The curvature expression (11.1) was derived by the transport of a vector field W along a closed parallelogram, constructed by two other independent vector fields U and V and their parallel transports as described in Fig. 2.5. This construction is independent of the choice of the space where the operator acts and it is independent of a previous choice of the connection. Finally it is independent of the choice of a basis in that space. As we said before, the Riemann curvature was conceived at the time when Riemann was defining also his metric geometry, so that the association of the above expression with a metric connection is very common, albeit unnecessary. As evidenced by the Yang–Mills theory, the same expression holds for all other fundamental interactions, when the symmetry of the field is defined. This result is supported by a vast experimental background.

M.D. Maia, *Geometry of the Fundamental Interactions*,
DOI 10.1007/978-1-4419-8273-5_11,

In particular, considering the tangent spaces to a Riemannian manifold and choosing a tangent basis $\{e_\mu\}$, the Riemann curvature tensor (11.1) is usually expressed by its components in that basis as

$$\mathscr{R}(e_\mu, e_\nu)e_\rho = R_{\mu\nu\rho}{}^{\sigma} e_\sigma = [\nabla_\mu, \nabla_\nu]e_\rho \tag{11.2}$$

where we simplified the notation $\nabla_\mu = \nabla_{e_\mu}$.

In the Riemannian geometry, the association of the connection with a metric $g_{\mu\nu}$ is made with the additional assumption that $g_{\mu\nu;p} = 0$, called the metricity condition as explained in Chapter 2. By expressing this operation in terms of the Christoffel symbols $\nabla_\mu e_\nu = \Gamma^\rho_{\mu\nu} e_\rho$, after a cyclic permutation of the indices, summing the two first results, and subtracting the third, we obtain the Levi-Civita metric connection of Riemann's metric geometry [48]

$$\Gamma^\rho_{\mu\nu} = \frac{1}{2} g^{\rho\sigma}(g_{\mu\sigma,\nu} + g_{\nu\sigma,\mu} - g_{\mu\nu,\sigma}) \tag{11.3}$$

The Riemann tensor derived from the above connection was applied by Einstein in 1916 to obtain the gravitational field equations. These equations were strongly motivated by Newton's gravitational theory, which as seen in Chapter 5 can also be formulated in geometrical terms. This geometrical interpretations of Newton's gravitation led to a simple derivation of Einstein's equations, which is now derived from the Einstein–Hilbert action principle, with respect to the metric

$$\frac{\delta}{\delta g_{\mu\nu}} \int R\sqrt{-g}dv = 0 \tag{11.4}$$

where R denotes the Ricci scalar $R = g^{\mu\rho} g^{\nu\sigma} R_{\mu\nu\rho\sigma}$.

Although this is seldom mentioned, we find it relevant in the present discussion to assign a geometrical interpretation to the Einstein–Hilbert action principle: Since the curvature scalar R is the simplest scalar term directly derived from the Riemann tensor, then *(11.4) has the meaning that the gravitational field follows from the smoothest possible geometry,* in the sense that it corresponds to the smallest possible variations of the Riemann curvature.

In relativistic cosmology the above interpretation of the variational principle can be rephrased a la Leibniz, by saying *the universe where we live is the smoothest among all possible universes.*

There is an alternative procedure to the variation of (11.4) which is called the *Palatini formulation* of Einstein's gravitation. In this formulation, the connection and the metric are initially considered to be two independent variables. Only after the metricity condition $\nabla g = 0$ is imposed we obtain the Levi-Civita connection [128]. The Palatini formulation has been applied in the comparison between gravitation and gauge theory [127].

The symmetry of Einstein's theory was chosen in a separate postulate to be the group of diffeomorphism of the space–time. That is, unlike the case of special relativity all coordinate systems of the space–time are equally good, meaning that no observer is preferred in detriment of another. Such democratic principle is often claimed to be a strong aspect of Einstein's theory (without any political connotations intended). On the other hand, the diffeomorphism invariance of the theory has been the source of the main difficulties of Einstein's theory, mainly with respect to a quantum theory of gravity. Indeed, the diffeomorphism group is a Lie group with infinite dimensions, whose unitary irreducible representations required in quantum gravity are very difficult if not impossible to classify.

As we have seen in Chapter 9, the concept of curvature in gauge theories is the same as in (11.1). The difference with general relativity resides in the choice of the connection. Following the example of the electromagnetic field, Yang and Mills proposed that the field strengths of the gauge fields are given by the curvature

$$F(U,V) = [D_U, D_V] \tag{11.5}$$

where now $D_\mu = I\partial_\mu + A_\mu$ is the covariant derivative associated with the connection A_μ.

Thus, different from the postulated Levi-Civita metric connection of general relativity, where the metric is determined by solving Einstein's equations, in gauge theory *the gauge connection is determined as a solution of the gauge field equations* resulting from the variational principle

$$\frac{\delta}{\delta A_\mu}\int \frac{1}{4} trF_{\mu\nu}F^{\mu\nu}dv = 0 \tag{11.6}$$

Therefore, the single idea of Riemann curvature has provided us with two successful variational principles (11.4) and (11.6), which are able to describe all known fundamental interactions and have predicted new results. The question that remains is, if we take the Riemann curvature as a fundamentally proved concept, *why do we need two different variational principles?* Can gravitation be derived from (11.6) or can all gauge fields be derived from (11.4)? In the search for a unity of physics, both possibilities have been tried, but they are not conclusive. New alternatives based on entirely different principles have also been proposed, sometimes even dispensing with the whole concept of continuity, fields, and interactions. In the following we will discuss, even if briefly, some of the above questions on the unity of the action principles involving the Riemann concept of curvature [129].

11.2 Gauge Gravity

The suggestion that gravitation based on Einstein's theory or a slight modification of it can be described as a gauge theory appeared around 1960, when various authors proposed the use of the Poincaré group of the tangent Minkowski's space–times as the gauge group of gravitation. That suggestion leads to the emergence of torsion in space–time, associated with the translational subgroup of the Poincaré group. The physical meaning of that torsion remains unclear.

On the other hand, to be a gauge theory the Einstein variational principle should be changed to something like in (11.5) [130]. This means a really fundamental modification of Einstein's theory.

In more general terms, back in 1962 Cornelius Lanczos conjectured that Einstein's gravitation could have properties similar to gauge theories, including the dualities of the Riemann tensor. That is, the four-index Riemann tensor components would be derived from a *three-index connection* $A_{\alpha\beta\gamma}$ [131]. This has led to an intense search for such three index connections. In spite of the many efforts, the existence of such potential was shown only in particular instances [132–135].

Clearly, if we wish to compare any two theories they must be written in the same mathematical language. In this sense, to describe gravitation as a gauge theory gravitation must be written in the language of gauge theory and not vice versa. This implies that such gauge theory of gravitation must have a Yang–Mills type Lagrangian, written in terms of the curvature operators. In this case, the appropriate procedure would be like in the standard Yang–Mills theory: identify a local gauge symmetry of gravitation as a Lie group; write its Lie algebra; write the corresponding curvature operator; and finally write the Yang–Mills equations. Only then we solve these equations to determine the gauge connection.

From these general lines we may conclude that *Einstein's theory of gravitation is not a gauge theory of gravitation*, for several reasons: To start with, the Einstein–Hilbert Lagrangian (11.1) is linear on the Riemann curvature tensor, whereas in gauge theory the Lagrangian is quadratic in the Riemann curvature tensor as (11.5). Second, the diffeomorphism invariance of the base space (the space–time) may interfere with the determination of the connection expressed in terms of dx^μ. However, nothing prevents the construction of an alternative gauge theory of gravitation, as long as it is motivated by the solutions of the current problems of gravitation: That is, hopefully quantizable in the sense described by 'tHooft [41] and hopefully being capable of explaining the 96% of the gravitational field of the universe. In the following we will describe the general lines of what the gauge theory of gravitation should look like, including the possible gauge symmetries.

We start with a principal bundle $(\mathscr{M}, G, \pi, T\mathscr{B})$ of a Lie symmetry group G for a Yang–Mills type Lagrangian constructed with the Riemann curvature of the manifold acting on a total space $T\mathscr{B}$, defined on a four-dimensional space–time manifold $\mathscr{M}$. As in the general case, the principal bundle is trivialized by the dual of the adjoint representation of the Lie algebra $\mathscr{M} \times \mathscr{G}^*$.

The Yang–Mills gravitational Lagrangian for the Riemann curvature (11.1) should be like in all other Yang–Mills fields

$$\mathscr{L} = \frac{1}{4} tr \mathscr{R}_{\mu\nu} \mathscr{R}^{\mu\nu} \tag{11.7}$$

where $\mathscr{R} = [D_\mu, D_\nu]$ and $D_\mu = I\partial_\mu + A_\mu$. The connection A_μ is to be determined by solving the Yang–Mills equations resulting from the above Lagrangian.

In practice we need to write the components of (11.7) in the basis $\{e_\mu\}$ of the dual Lie algebra $\mathscr{G}^*$ of the symmetry group G (not yet defined) to obtain a vector-matrix in the Lie algebra. Then the Yang–Mills Lagrangian for the gravitational field would be:

$$\mathscr{R}^{\mu\nu}\mathscr{R}_{\mu\nu}(e_\rho) = R_{\mu\nu\rho}{}^{\sigma}\mathscr{R}^{\mu\nu}(e_\sigma) = R_{\mu\nu\rho}{}^{\sigma}R^{\mu\nu}{}_{\sigma\tau}\,e^\tau = R_{\mu\nu\rho\sigma}R^{\mu\nu\sigma\tau}e_\tau$$

Taking its trace and replacing in (11.7) we obtain the Yang–Mills gravitational Lagrangian, for an yet unspecified gauge symmetry

$$\mathscr{L} = \frac{1}{4}trR_{\mu\nu\rho\sigma}R^{\mu\nu\rho\sigma} \tag{11.8}$$

The inclusion of $\sqrt{-g}$ is optional, depending on whether the diffeomorphism invariance is postulated or not.

The above Lagrangian looks like one of the so-called $f(R)$ (quadratic) gravitational theories, defined by a Lagrangian like $\mathscr{L} = f(R)\sqrt{g}$, where f is an arbitrary function of scalars built with the Riemann tensor. The field equations are the same Yang–Mills equations.

The local gauge symmetry of the Lagrangian (11.8) cannot be arbitrarily chosen. This is dictated by the necessity to have special relativity as the limit of vanishing gravitation. Furthermore, the chosen gauge symmetry must be able to mix with the other gauge groups. Here we discuss only three examples:

The simplest choice of symmetry is a local Lorentz symmetry acting on each tangent fiber of the space–time, whose connection is defined by the solutions of the field equations derived from (11.7). Its Lie algebra is well established and we may readily construct its adjoint representation (which is locally defined).

Local Lorentz gauge symmetry has been suggested by several authors (see, e.g., [136, 137]). The most attractive property of one such theory is that it is compatible with the combination with the other gauge symmetries in the standard model, which becomes

$$\text{Lorentz}_{\text{local}} \times U(1) \times SU(2) \times SU(3)$$

There are two good arguments to exclude the complete Poincaré group as a gauge symmetry. One of them is that the Lie algebra operators of the translations always commute, so that the curvature operators associated with these translations vanish, and they do not contribute to the Lagrangian. However, the translations can be associated with torsion, which lead either to the Einstein–Cartan theory or to a pure torsion theory. In such case, we need to reinterpret gauge theory also as a torsion theory.

The second reason to avoid the Poincaré symmetry is the mixing symmetry problem mentioned in the introduction: If we combine the Poincaré group with the other gauge symmetries like in the above Cartesian product, we end up having all particles belonging to the same spin multiplet, also with the same mass, which is not true. This is again a consequence of the translational subgroup of the Poincaré group. Therefore the inclusion of the complete Poincaré symmetry as the gauge group of gravitation will not describe the known spectrum of particle mass and spins.

One possible way out of this difficulty is to replace the Poincaré group by the deSitter symmetry in four-dimensional space–time with a constant curvature Λ. By a process known as the Inonu–Wigner *group contraction*, we obtain the Poincaré group when $\Lambda \rightarrow 0$ [138].

Another possibility to incorporate the translational symmetry as part of the local gauge symmetry is to consider the 15-parameter local *conformal group* in each tangent space–time [35]. The Yang–Mills equations are invariant under the conformal group and the whole Poincaré group is a subgroup of it.

The conformal group is isomorphic to the pseudo-orthogonal group $SO(4, 2)$ and the representations of it have been classified [139]. The group $SO(4, 2)$ can be interpreted as the group of isometries in a six-dimensional space with two time-like dimensions.

Since the Yang–Mills equations are consistent only in four dimensions, the only way to write the Yang–Mills equations in a six-dimensional space is to have the four-dimensional space–time as a subspace embedded in that space.

The conformal invariance of Maxwell's equations was discovered in 1909/1910 [74, 75]. However, two factors have contributed to its early dismissal: One was the fact that it is not causal. That means that we need to consider the advanced component of the electromagnetic potential, something that was not understood, and perhaps it is not yet understood today, even considering Feynman's discussion on that subject in 1942 [140]. The other factor was that its association with a six-dimensional space–time, with two times, was not really very attractive in a period when almost everyone thought that three dimensions were enough.

Until recently causality was one stronghold against speculative theories [76]. However, at the quantum level it may appear different. The use of the conformal group has been recently explored as a way to study quantum fields in presence of the gravitational field. It originated with a conjecture known as the ADS/CFT correspondence, by means of which the quantization of the standard gauge theories in Minkowski space–time can be transferred to a curved space–time, so that eventually it induces quantum fluctuations of the metric geometry [36]. The ADS stands for the anti-deSitter space–time which is a four-dimensional subspace of $M_6(4, 2)$. In this particular line of thought we may consider the universal covering group of the conformal group which is $SU(2, 2)_{\text{local}}$, also known as the *twistor* group or the local gauge group, suggesting the extended gauge unification to

$$SU(2, 2)_{\text{local}} \times U(1) \times SU(2) \times SU(3)$$

where hopefully the implications of the translational subgroup will not show up. We should note in passing that the same twistor group has been present in one recent revision of the string program [141].

11.3 Loop Gravity

One interesting question is, why was the Riemann curvature defined by transport of a vector field W along a four-sided parallelogram? Could it be derived by a three-sided figure, or perhaps an infinite number of small line segments? What prevents us considering the definition of the same concept using just the transport of W along a single continuous loop? Actually nothing, but the curvature operator will depend on the loop path followed by W. The expression of the Riemann curvature in a closed loop γ was evaluated by Wilson in 1973 for a three-dimensional surface in a four-dimensional space–time [142].

Consider the transport of a tangent vector field along a single continuous loop γ, starting and ending at point p. As in Fig. 2.5, the end vector W' does not coincide with the original vector W. Their difference can be expressed by the loop integral [142, 143]

$$R(\Gamma) = P e^{\oint_\gamma \Gamma_i dx^i}, \quad i = 1, \ldots, 3 \tag{11.9}$$

where the coefficients Γ_i are the components of the affine connection ∇ evaluated in a base of the three-dimensional space (a triad, in the used language). P is an ordering fact which depends on the path integration.

It follows that the circular motion of the basis characterizes a local group, called *the group of holonomy of the triad*. This is a Lie group with three parameters, so that it is isomorphic to $SO(3)$, which in turn is isomorphic to $SU(2)$. The *ordering factor P* is defined by the orientation of the transport along the closed curve.

Admitting that Γ_i are continuous functions of the coordinates along the loop, the exponential in (11.9) is a well-defined real analytic function represented by the standard exponential converging positive power series

$$e^{\oint_\gamma \Gamma_i dx^i} = 1 + \oint_\gamma \Gamma_i dx^i + \left(\oint_\gamma \Gamma_i dx^i \right)^2 + \cdots \tag{11.10}$$

Therefore, (11.9) is a well-defined function of the connection Γ_i, describing the classical curvature of the three-dimensional surface in the sense of Riemann.

The quantum version of (11.9) can be obtained by applying the classical quantum correspondence

$$\Gamma_i \longleftrightarrow i\hbar \hat{\mathscr{A}}_i, \quad i = 1, \ldots, 3 \tag{11.11}$$

where $\hat{\mathscr{A}}_i$ denote the three-dimensional components of the quantized $SU(2)$ gauge potential. As we have seen in the discussion of the Yang–Mills theory in the previous section, the Lie algebra of $SU(2)$ can be written in terms of the Pauli matrices as $\Gamma = d\Gamma_\mu \sigma^\mu$, so that the connection one-form can be expressed as

$$\hat{\mathscr{A}}_\mu dx^\mu = d\Gamma_\mu \sigma^\mu$$

In this representation $\hat{\mathscr{A}}_\mu$ are quaternion fields. Thus, the three-dimensional curvature operator (11.9) leads to the *quantum curvature* operator of the three-dimensional hypersurface

$$\hat{\mathscr{R}}(\mathscr{A}) = \mathscr{P}e^{i\hbar \oint_\gamma \hat{\mathscr{A}}_i dx^i}, \quad i = 1, \ldots, 3 \tag{11.12}$$

where the three-dimensional components $\hat{\mathscr{A}}_i$ are obtained from the imaginary (vector) part of the full quaternion $\hat{\mathscr{A}}_\mu$ [118].

Hence the quantum fluctuations of the $SU(2)$ gauge field $\hat{\mathscr{A}}_i$ in the exponential function induce the quantum fluctuations of the three-dimensional curvature and consequently the quantum fluctuations of the gravitational field [144].

The difficulty facing (11.12) is the same as we discussed in the quaternion representation of instantons in the previous chapter. By the same token, the analytical properties of the quaternion exponential of a quaternion function are not well defined to be rightfully represented by a convergence of positive power series. Therefore, to proceed with this interesting proposal for quantum gravity, either we define analytical quaternion functions, including the quaternion exponential by the power series, or else we reinvent the integral (11.12) as a limit of a discrete curvature, which means a discrete geometry. The latter option caused a change of course in the original induced quantization program of Ashtekar, where the integral could be derived from a discrete space–time structure [145, 146].

Quite justifiable this later phase of loop quantum gravity has been implemented by an earlier discrete space–time structure described by Penrose, called *Spin Network*. In essence, this is a projective space where the projective rays (or lines) are associated with the possible eigenvalues of the spin Casimir operator of the Lorentz group [147].

11.4 Deformable Gravity

Since all field strengths in gauge and gravitation use the same concept of curvature, we infer that the importance of Riemann curvature to the fundamental interactions is enormous. However, the local shape of Riemannian space–time is not uniquely determined by the Riemann tensor. So, the observed fundamental interactions do not correspond to a specific geometry, but to a class of equivalence of geometries. The ambiguity of the Riemann curvature to determine the shape was noted by Riemann in his original paper: ...*We may, however, abstract from external relations by considering deformations which leave the lengths of lines within the surfaces unaltered, i. e, by considering arbitrary bendings – without stretching – of such surfaces, and by regarding all surfaces obtained from one another in this way as equivalent. Thus, for example, arbitrary cylindrical or conical surfaces count as equivalent to a plane...*[5].

Does this difference in shape make a difference to physics? From the point of view of gauge theory, the shape difference seems to be less critical. After all, the connections are determined *from* the Riemann curvature and so, regardless of any possible ambiguity that the curvature may have, the experimental evidences in high energy physics give a solid support for the standard gauge theory. On the other hand, in Einstein's gravitation the connection is postulated and the observables of the gravitational field are determined by the eigenvalues of the Riemann tensor. As we have already commented the experimental tests for the gravitational interaction in cosmology do not correspond to gravitation described by Einstein's equations.

The problem of the *shape ambiguity of Riemann tensor* appeared in the early days of the Riemann geometry, long before it was used to describe the gravitational field. The concern then had to do with the lack of intuitiveness of the Riemann tensor, which should convey the idea of shape of the manifold. In particular for a physical manifold such shape should be an observable.

The earliest known proposition to solve this problem was given by Luigi Schlaefli in 1871, when he conjectured that the shape of a Riemannian manifold could be defined *if the Riemannian manifolds could be embedded in a 1:1 fashion into a higher-dimensional space* [148]. In this way, the Riemann curvature of the space–time can be compared with the Riemann curvature of the embedding space, such that their difference defines the shape in terms of the extrinsic curvature. Since the embedding is a 1:1 map, after the characterization of the shape we could reverse the embedding and return to the intrinsic geometry, but now knowing about the shape defined by the extrinsic curvature, for example, telling the difference between a plane and a cylinder. This would be like some kind of fine-tuning the Riemann curvature or, if you like, the gravitational field.

Therefore, Schlaefli's conjecture provided the definitive solution for the shape problem of the Riemann geometry, but it was not so easy to implement because it depended on the development and solutions of the Gauss–Codazzi–Ricci equations. These are non-linear equations involving the metric, $g_{\mu\nu}$, the extrinsic curvature $k_{a\mu\nu}$, and the third fundamental form $A_{\mu ab}$ as independent variables. They are the integrability conditions for the existence of the embedding [48].

Earlier theorems on the solutions of the Gauss–Codazzi–Ricci equations made explicit use of the expansion of the embedding functions in positive power series (analytic functions) [149, 150]. The general solution of the local embedding problem based on differentiable functions was derived by John Nash in 1956, using a process of *smooth deformation* of Riemannian manifolds [151].

Nash's Local Embedding for Space–Times

Nash's theorem was originally shown for positive definite metrics but it was soon extended to non-positive metrics. The smoothing process is generally considered to be difficult to understand and we will present an alternative and much simpler derivation of the Nash geometric flow condition. The theorem can be announced as:

Given a space–time $\bar{\mathcal{M}}_4$, *such that its isometric local embedding in a larger Riemannian manifold* $\mathcal{V}_D$, *by a regular and differentiable map* $X : \bar{\mathcal{M}} \to \mathcal{V}_D$, *is known. Then it is possible to smoothly deform the metric of* $\bar{\mathcal{M}}$ *along the extra dimensions to obtain another embedded geometry space–time. Using the inverse embedding map, the deformed space–time may be removed from the embedding to obtain the intrinsic four-dimensional deformed space–time.*

For simplicity we will show the result for only one extra dimension, so that all deformations are limited to fit into the same five-dimensional space. The theorem holds for any number of dimensions and metric signatures, but here we set $D = 5$. It is interesting to notice that the signature of the embedding space is not open to choice [152] the signature of the extra dimensions is not open to choice, but they are determined by the embedding equations.

Denote the metric components of the embedding space by G_{AB}, the metric of $\mathscr{M}$ by $\bar{g}_{\mu\nu}$ both in arbitrary coordinates. The embedding map is a vector of V_D, with components X^A. The unit vector orthogonal to the embedded geometry has components η^A. Then the isometric embedding is defined by

$$X^A{}_{,\mu}X^B{}_{,\nu}G_{AB} = \bar{g}_{\mu\nu},\ X^A{}_{,\mu}\bar{\eta}^B G_{AB} = 0,\ \bar{\eta}^A\bar{\eta}^B G_{AB} = 1 \tag{11.13}$$

The extrinsic curvature of $\bar{V}_n$ is by definition the projection of the variation of η on the tangent plane and is given by [48]

$$\bar{k}_{\mu\nu} = -\bar{X}^A{}_{,\mu}\bar{\eta}^B{}_{,\nu}G_{AB} = X^A{}_{,\mu\nu}\bar{\eta}^B G_{AB} \tag{11.14}$$

Consider the one-parameter group of diffeomorphisms defined in V_D, applied to points in $\mathscr{M}_4$ by a map

$$h_y(p) : V_D \rightarrow V_D$$

defining a curve in V_D, called the orbit of p, with parameter y, passing through p, and with tangent vector at p given by the orthogonal unit vector:

$$\alpha(y) = h_y(p),\quad \alpha'(p) = \eta(p)$$

From the fundamental theory of curves we know that such a curve exists and it is unique.

The group properties are characterized by the composition map defined by

$$h_y o h_{y'}(p) = h_{y+y'}(p),\quad h_0(p) = p,\quad h_{y-y} = h_0(p) = p$$

Applying such diffeomorphism to all points in a neighborhood $\cup_p$ of p, we obtain a local congruence of orbits in $\mathscr{V}_D$, each orbit with its own parameter y. The resulting set of points is an aleatory distribution, not necessarily characterizing an embedded submanifold. For this we need further conditions.

Given a geometric object $\bar{\Omega}$ at a point in $\bar{\mathscr{M}}$, the Lie transport of $\bar{\Omega}$ for a distance δy along the orbit produces a new object given by [81]

$$\Omega = \bar{\Omega} + \delta y \pounds_\eta \bar{\Omega}$$

where $\pounds_\eta$ denotes the Lie derivative of $\bar{\Omega}_p$ with respect to η.

In particular, the Lie transport of the Gaussian frame $\{X^A_\mu, \bar{\eta}^A_a\}$ of $\mathscr{M}$ gives

$$Z^A{}_{,\mu} = X^A{}_{,\mu} + \delta y\; \pounds_\eta X^A{}_{,\mu} = X^A{}_{,\mu} + \delta y\; \eta^A{}_{,\mu} \tag{11.15}$$

$$\eta^A = \bar{\eta}^A + \delta y\; [\bar{\eta}, \bar{\eta}]^A \quad = \; \bar{\eta}^A \tag{11.16}$$

The set of coordinates Z^A obtained by integrating these equations may describe another embedded differentiable manifold in V_D only if they satisfy the new embedding equations

$$Z^A{}_{,\mu} Z^B{}_{,\nu} G_{AB} = g_{\mu\nu}, \; Z^A{}_{,\mu} \eta^B G_{AB} = 0, \; \eta^A \eta^B G_{AB} = 1 \tag{11.17}$$

Replacing (11.15) and (11.16) in (11.17) and using definition (11.14) we obtain the new components

$$g_{\mu\nu} = \bar{g}_{\mu\nu} - 2y\bar{k}_{\mu\nu} + y^2 \bar{g}^{\rho\sigma} \bar{k}_{\mu\rho} \bar{k}_{\nu\sigma} \tag{11.18}$$

$$k_{\mu\nu} = \bar{k}_{\mu\nu} - 2y \bar{g}^{\rho\sigma} \bar{k}_{\mu\rho} \bar{k}_{\nu\sigma} \tag{11.19}$$

and taking the derivative of (11.18) with respect to y and comparing with (11.19) we obtain Nash's geometric flow condition

$$k_{\mu\nu} = -\frac{1}{2} \frac{\partial g_{\mu\nu}}{\partial y} \tag{11.20}$$

Actually (11.20) has been in use in general relativity since 1971, under the designation of *York relation* in the study of initial value condition in the 3+1 decomposition of space–time when y is taken as the time variable[1] [155].

To complete Nash's theorem we require that (11.17) be integrated. Only then Z^A will describe the embedding map of the *deformed Riemannian manifold* $\mathscr{M}_4$. This is obtained by solving the integrability conditions for (11.17). They are the Gauss–Codazzi equations (In the case of just one extra dimension the Ricci equation does not exist.) obtained from the components of the Riemann tensor ${}^5R_{ABCD}$ of V_D, in the Lie transported Gaussian frame:

[1] Although the idea of smooth deformation of a Riemannian geometry is older, it has become popular only recently in the form of $R_{\mu\nu} = -\frac{1}{2}\frac{\partial g_{\mu\nu}}{\partial y}$ where $R_{\mu\nu}$ is the Ricci tensor and y represents any coordinate in the manifold. This expression was inspired by the Fourier law on heat flow and was successfully applied to solve the Poincaré conjecture on the continuous deformation of a compact three-dimensional manifold into a sphere [153, 154]. Unfortunately the Ricci flow is non-relativistic and it is not compatible with Einstein's gravitation in four dimensions.

$$ {}^5R_{ABCD}Z^A{}_{,\alpha}Z^B{}_{,\beta}Z^C{}_{,\gamma}Z^D{}_{,\delta} = R_{\alpha\beta\gamma\delta} + (k_{\alpha\gamma}k_{\beta\delta} - k_{\alpha\delta}k_{\beta\gamma}) \tag{11.21} $$

$$ {}^5R_{ABCD}Z^A{}_{,\alpha}Z^B{}_{,\beta}Z^C{}_{,\gamma}\eta^D = k_{\alpha[\beta;\gamma]} \tag{11.22} $$

Considering that the embedding of the original manifold $\mathcal{M}$ satisfies the Gauss–Codazzi equations, it follows that the above equations are automatically satisfied when we apply (11.15) and (11.16). Then using the the Weingarten equation we may finally determine Z. It is important to observe that Z remains a regular function so that the local inverse of Z determines the new and deformed [48].

From the first of these equations (Gauss' equation) we clearly see how Schlaefli's solution of the shape ambiguity in Riemann's tensor is solved: *The Riemann tensor of the embedded manifold is compared with the Riemann tensor of the host space*, and their difference is given by the extrinsic curvature. The second equation defines a condition on the extrinsic curvature in terms of other components of the Riemann tensor of the embedding space.

Deformable Gravity

Applying the above deformation condition to a space–time of general relativity, obtain new Einstein's equations in four dimensions describing a deformed gravity as follows.

The physical interpretation of (11.20) is that the gravitational field represented by the metric of space–time propagates along the extra dimensions of the embedding space as well as propagating in the space–time itself in accordance with Einstein's equations. However, the propagation along the extra dimensions is given by the extrinsic curvature. Clearly these two different forms of propagation must be consistent. This is obtained by a definition of the geometry of the embedding space. Since the embedding is isometric, the deformed metric is induced by the metric of the embedding space. Therefore, the metric geometry of the embedding space is also derived from the same Einstein–Hilbert principle, leading to the higher dimensional Einstein's equations. In the case of five dimensions these equations are

$$ {}^5R_{AB} - \frac{1}{2}\,{}^5R\mathcal{G}_{AB} = G_* T^*_{AB} \tag{11.23} $$

where G_* is a new gravitational constant compatible with the higher dimensional geometry [156] and where T^*_{AB} denotes the components of the energy–momentum tensor of the known material sources capable of interacting with the electromagnetic and nuclear forces. As we have seen in Chapter 10, these are composed of the ordinary matter and gauge fields consistently defined and observed in the four-dimensional space–times only. Thus, in order to reproduce the 4-dimensional Einstein's equations, this corresponds to say that the projected components of the right hand side of (7.1) are

$$G_* Z^A_{,\mu} Z^B_{,\nu} T^*_{AB} = 8\pi G T_{\mu\nu}$$
$$Z^A_{,\mu} \eta^B T^*_{AB} = 0$$
$$\eta^A \eta^B T^*_{AB} = 0$$

Therefore the projected components of (11.23) gives the four-dimensional Einstein's equations for the deformed space–time

$$R_{\mu\nu} - \frac{1}{2} R g_{\mu\nu} - Q_{\mu\nu} = 8\pi G T_{\mu\nu} \tag{11.24}$$

$$k_{\mu;\rho}^{\rho} - h_{,\mu} = 0 \tag{11.25}$$

where we have denoted the mean curvature of the space–time by $h = \sqrt{g^{\mu\nu} k_{\mu\nu}}$, Gaussian curvature by $K = \sqrt{k^{\mu\nu} k_{\mu\nu}}$, and

$$Q_{\mu\nu} = g^{\rho\sigma} k_{\mu\rho} k_{\nu\sigma} - k_{\mu\nu} h - \frac{1}{2}\left(K^2 - h^2\right) g_{\mu\nu} \tag{11.26}$$

The tensor $Q_{\mu\nu}$, called the deformation tensor does not appear in the usual Einstein's equations. It is conserved in the sense that

$$Q^{\mu\nu}{}_{;\nu} = 0 \tag{11.27}$$

so that it is an observable in space–time representing the missing shape information, under the conditions of Noether's theorem described at the end of Chapter 8. In other words, *there are observables effects in space–time associated with the extrinsic curvature.*

11.5 Kaluza–Klein Gravity

The success of Einstein's theory of geometrical gravitation motivated the possibility that the original Riemannian paradigm should be modified to include the other three fundamental interactions. This would be a reversal of the current trend to define a gauge theory of gravitation. In the Kaluza–Klein program all gauge interactions are contained in the same Einstein–Hilbert principle, but applied to a higher dimensional Riemannian manifold.

Kaluza–Klein theory was defined in 1920 by Theodore Kaluza as a five-dimensional theory based on the Einstein–Hilbert principle, proposing a unification of the electromagnetic and gravitational fields. The theory was made more consistent by Felix Klein in 1926. Much later in 1963 Kaluza–Klein theory was generalized to include all gauge fields [157].

The theory assumes that the physical space is a Riemannian manifold with a product topology $V_4 \times B_N$, where V_4 denotes a space–time of general relativity and

B_N is an *N–dimensional* compact space. The geometry of the total space was also defined by Einstein–Hilbert in $4 + N$ dimensions. Only in 1984 it was understood that the Kaluza–Klein theory was not able to reproduce the observations at the level of the electroweak theory.

Essentially, the problem of the proposed theory resided in the compact internal space B_N which was proposed to have diameter equal to Planck's length, 10^{-33}cm. This would guarantee that this space would not be "visible" by any known gauge probes. However, when evaluating the behavior of fermions described by the higher dimensional Dirac equation, it was found that at the lower energy limit of the theory, at the electroweak scale, the fermion mass term generated by the internal components would not go away, perturbing the observed behavior (the chirality) of fermions at the electroweak level [158]. Within the assumed postulates of the theory, mainly the product topology, there was nothing to be done. A number of schemes were proposed to save the theory, but after 20 years of hard work the theory was abandoned by 1985 (see, e.g., [159, 160]).

References

1. F. Zwicky, Helv. Phys. Acta **6**, 110 (1933).
2. V. Rubin & W. K. Ford Jr., Astrophys. J. **159**, 379 (1970), doi:10.1086/150317.
3. S. Perlmutter et al. (The Supernova Cosmology Project), Astrophys. J. **517**, 56586, doi:10.1086/307221.
4. N. Cantor, *In the Wake of the Plague*, Simon and Schuster, p. 112 UK (2002).
5. B. Riemann, *Ueber die Hypothesen, welche der Geometrie zu Grunde liegen*, pp. 272–287, Gesammelte Matematische Werke, Leipzig (1892). Tranlation by W. D. Clifford, On the hypothesis which lie at the base of geometry, Nature **8**, 114–117 and 136–137 (1873).
6. P. Pesic, *Beyond Geometry*, Dover, New York, NY (2007).
7. R. Penrose, *The Road to Reality*, A. Knopf, New York, NY (2004).
8. L. O'Raifeartaigh & N. Straumann, hep-ph/9810254 (1998).
9. H. Weyl, *Space Time Matter*, 4th edition, Dover, New York, NY (1922).
10. E. Noether, *Invariante Variationsprobleme*. Nachr. D. König. Gesellsch. D. Wiss. Zu Göttingen, Math-phys. 235, Klasse (1918).
11. J. D. Jackson, *Electrodynamics*, Willey, New York, NY sixth print (1967).
12. V. Fock, Zeit. Phys. **39**, 226 (1927).
13. F. London, Zeit. Phys. **42**, 375 (1927).
14. H. Weyl, Zeit. Phys. **56**, 330 (1929).
15. J. Schwinger, *Quantum Electrodynamics*, Dover, New York, NY (1958).
16. Y. Aharonov & D. Bohm, Phys. Rev. **115**, 485 (1959).
17. A. Actor, Rev. Mod. Phys. **51**, 461 (1979).
18. K. Moriyasu, *An Elementary Primer for Gauge Theory*, World Scientific, Singapura (1983).
19. A. Pickering, *Constructing Quarks*, University of Chicago Press, Chicago (1984).
20. W. Heisenberg, Zeits. Phys. **77**, 1 (1932).
21. E. Wigner, Phys. Rev. **51**, 106 (1937).
22. C. N. Yang & R. Mills, Phys. Rev. **96**, 191 (1954), doi:10.1103/PhysRev.96.191.
23. Sheldon Glashow, Rev. Mod. Physics, **52**, Issue 3: 539 (1980).
24. S. Weinberg, Phys. Rev. Lett. **19**, 1264 (1967).
25. A. Salam, Proceedings of the 8th Nobel Simposium, Almquist Forlag, Stockholm (1968).
26. James Joyce, *Finnegans Wake*, Faber and Faber, London (1939).
27. M. Gell-Mann, *The Eightfold Way: A Theory of Strong Interaction Symmetry*, In Synchroton Laboratory Report CTSL-20 (California Institute of Technology), (1961), M. Gell-Mann & Y. Ne'emann (Eds.), The Eightfold Way, Benjamin, New York, NY (1964).
28. Y. Ne'emann, Nuclear Phys. **26**, 222 (1961).
29. R. Slansky, Phys. Rep. **79**, 1 (1981).
30. Eugene Wigner, Ann. Math. **40**, 149 (1939).
31. L. O'Raifeartaigh, Phys. Rev. **139B**, 1052 (1965).
32. Sidney Coleman & Jeffrey Mandula, Phys. Rev. **159**, 1251 (1967), doi:10.1103/PhysRev.159.1251.

M.D. Maia, *Geometry of the Fundamental Interactions*,
DOI 10.1007/978-1-4419-8273-5,

33. M. Flato & D. Sternheimer, Phys. Rev. Lett. **15**, 934 (1965).
34. J. J. Aghassi, P. Roman, & R. M. Santilli, J. Math. Phys. **11**, 2297 (1970), doi:10.1063/1.1665396.
35. R. Penrose, J. Math. Phys. **8**, 345 (1967).
36. Juan M. Maldacena, Adv. Theor. Math. Phys. **2**, 231 (1998).
37. T. Wess & B. Zumino, Nuclear Phys. **B70** (1974).
38. P. Higgs, Phys. Rev. Lett. **12**, 132 (1964).
39. T. B. Kibble, Phys. Rev. Lett. **13**, 585 (1964).
40. M. Planck, *Theory of Heat Radiation*, 2nd edition, Blackiston's Son & Co, Philadelphia, PA (1907).
41. G. 't Hooft, Nuclear Phys. **B35**, 1967 (1971).
42. Immanuel Kant, *Critique of Pure Reason* (first published in 1781), English translation by J. M. D. Meiklejohn (1900), Chapter 2, section 2, p. 75, Dover, New York, NY (2003).
43. Michael Spivak, *A Comprehensive Introduction to Differential Geometry*, Vol. 2, p. 135, Publish or Perish, Delaware (1975).
44. Leopoldo Nachbin, *Topology and Order*, Van Nostrand, Princeton, NJ (1965).
45. Louis Auslander & Robert E. Mackenzie, *Introduction to Differentiable Manifolds*, Dover, New York, NY (1963).
46. William M. Boothby, *Introductions to Differentiable Manifolds and Riemannian Geometry*, Academic, New York, NY (1975).
47. Noel J. Hicks, *Notes on Differential Geometry*, Van Nostrand, Princeton, NJ (1965).
48. L. P. Eisenhart, *Riemannian Geometry*, Princeton University Press, Princeton, 6th print NJ, (1966).
49. Lee Smolin, *The Trouble with Physics*, Penguin, London (2006).
50. Ren Thom, *Structural Stability And Morphogenesis* , Westview Press, Boulder, CO (1994).
51. John A. Wheeler, *Geometrodynamics*, Academic, New York, NY (1962)
52. J. Ambjorn, J. Jurkiewicz, & R. Loll, Spin-06/16, ITP-UU-06/19, arXiv:hep-th/0604212v1 (2006).
53. Hermann Weyl, *The Continuum* (first published in 1918), Dover, New York, NY (1994).
54. Solomon Feferman, *The Significance of Hermann Weyl's Das Kontinuum*, In Proof Theory V. F. Hendricks et al. (Eds.), pp. 11–31, Kluwer/Springer, New York, NY (2000). see also http://math.stanford.edu/ feferman/papers/DasKontinuum.pdf
55. J. Goldstein, *Classical Mechanics*, Addison-Wesley, New York, NY (1980).
56. W. Elwood Byerly, *An Introduction to the use of Generalized Coordinates in Mechanics and Physics* (first published 1916), Dover, New York, NY (1965).
57. David Hilbert, *The Foundations of Geometry*, http://www.gutenberg.org/etext/17384.
58. Barrett, O'Neil, *Elementary Differential Geometry*, Academic, New York, NY (1996).
59. Hermann Weyl, *Symmetry*, Princeton University Press, Princeton, NJ (1952).
60. Sophus Lie & Georg Scheffers, *Vorlesungen über continuerliche Gruppen*, B. G. Teubner, Leipzig (1893).
61. Pierre Ramond, arXiv:hep-th/0301050v1 (2003).
62. V. S. Varadarajan, *Lie Groups, Lie Algebras, and Their Representations*, Prentice-Hall, Englewood Cliffs, NJ (1975).
63. N. Hammermesh, *Group Theory and Its Application to Physical Problems*, Dover, New York, NY (1962).
64. Gursey, *Introduction to Group Theory*, In Relativity, Groups and Topology, Proceedings of the Les Houches Summer School, C. DeWitt & B. DeWitt (Eds.), Gordon and Breach, New York (1963).
65. Wilhelm Killing, Math. Ann. **31**, 252–290 (1888).
66. A. S. Eddington, *Fundamental Theory*, Cambridge University Press, Cambridge (1948).
67. C. W. Kilmister, *Eddington's Search for a Fundamental Theory: A Key to the Universe*, Cambridge University Press, Cambridge (1994).

68. Tulio Levi-Civita, *The Absolute Differential Calculus. Calculus of Tensors* (first published 1926), Dover, New York, NY (2005).
69. Max Jammer, *Concepts of Space: The History of Theories of Space in Physics*, Harvard University Press, Cambridge, (MA) (1954).
70. C. Lanczos, *Space Through the Ages*, Academic, New York, NY (1970).
71. R. Penrose, *An Analysis of the Structure of Space-Time*. Adams Prize Essay, Cambridge University Press, Cambridge (1966).
72. Isaac Newton, *Philosophiae Naturalis Principia Mathematica*, S. Pepys, London (1686).
73. H. Minkowski, *The Principle of Relativity*, Dover, New York, NY (1952).
74. E. Cunningham, Proc. London Math. Soc. **8**, 77 (1909).
75. H. Bateman, Proc. London. Math. Soc. **8**, 223 (1910).
76. J. L. Synge, *Relativity Theory, vol 1, The Special Theory*, Van Nostrand, Princeton, NJ (1964).
77. S. Weinberg, Rev. Mod. Phys. **61**, 1 (1989).
78. Ray D'Inverno, *Introducing Einstein's Relativity*, The Clarendon Press, Oxford (1992).
79. K. Kuchar, *Time and Interpretations of Quantum Gravity*, In Proceedings of the 4th Canadian Conference on General Relativity and Relativistic Astrophysics, G. Kunstatter, D. Vincent, & J. Williams (Eds.), World Scientific, Singapore (1992).
80. M. C. Fernandes, C. Weinstein, & C. Blohmann, arXiv:1003.2857 (2010).
81. M. Crampin & F. A. E. Pirani, *Applicable Differential Geometry*, Cambridge University Press, Cambridge (1986).
82. L. D. Landau & E. Lifshitz, *The Classical Theory of Fields*, Pergamon, Oxford (1951).
83. Pierre Louis Moreau de Maupertuis, *Les loix du mouvement et du repos dduites d'un principe metaphysique*, In Accord de diffrentes loix de la nature qui avoient jusquici paru incompatibles, pp. 417–426, Histoire de l'Acadmie Royale des Sciences de Paris, 1744, (1748), http://gallica.bnf.fr/.
84. Leonhard Euler, *Methodus inveniendi Lineas Curvas* /Additamentum II. Marcus Michaelem Busquet, Geneve (1744).
85. Joseph-Louis Lagrange, *Mecanique Analytique, Courcier (1811)*, Cambridge University Press, Cambridge (2009).
86. V. I. Arnold, *Mathematical Methods of Classical Mechanics*, Springer, New York, NY (1989).
87. R. Abraham & J. E. Marsden, *Foundations of Mechanics*, AddisonWesley, Redwood City, CA (1987).
88. V. Mostepanenko et al., arXiv:0706.3283. Proceedings of the 10th Marcel Grossmann Meeting, Rio de janeiro (2003).
89. J. D. Bjorken & S. D. Drell, *Relativistic Quantum Fields* McGraw-Hill, New York, NY (1964).
90. J. Goldstone, Nuovo Cimento, **19**, 154 (1961).
91. J. Goldstone, A. Salam, & S. Weinberg, Phys. Rev. **127**, 965 (1962).
92. P. G. Drazin & R. S. Johnson, *Solitons an Introduction*, Cambridge University Press, Cambridge (1989).
93. John C. Slater & Nathaniel H. Franck, *Electromagnetism*, McGraw-Hill, New York, NY (1947).
94. James Clerk Maxwell, *A Treatise on Electricity and Magnetism* (first published 1873), Dover, New York, NY (1954).
95. H. B. Nielsen & P. Olesen, Nuclear Phys. **B61**, 45 (1973), doi:10.1016/0550-3213(73) 90350-7.
96. W. Meissner & R. Ochsenfeld, Naturwissenschaften **V21**, 787 (1933).
97. G. 't Hooft, Nuclear Phys. **B79**, 276 (1974).
98. A. M. Polyakov, Zh. Eksp. Teor. Fiz. Pis'ma. Red. **20**, 430 (1974) [JETP Lett. **20**, 194 (1974).
99. C. Chevalley, *The Construction and Study of Certain Important Algebras*, Mathematical Society, Japan (1955).

100. William R. Hamilton, *Lectures on Quaternions*, Royal Irish Academy (1853).
101. Wolfgang Pauli, Zeitschrift für Phisik **43**, 601 (1927), doi:10.1007/BF01397326.
102. Richard Brauer & Hermann Weyl, Am. J. Math. **57**, 425 (1935), doi:10.2307/2371218.
103. C. Chevalley, *The Algebraic Theory of Spinors and Clifford Algebras*, Columbia University Press, New York, NY (1954).
104. P. M. Dirac, Proc. R. Soc. London **A117**, 610 (1928).
105. L. Lederman & C. T. Hill, *Symmetry and the Beautiful Universe*, Prometheus Books, New York, NY (2004).
106. Yuval Ne'eman, *The Impact of Emmy Noether's Theorems on XX1st Century Physics*, pp. 83–101, Teicher (1999).
107. J. Leite Lopes, *Classical Symmetries*, Latin American School of Physics, Caracas (1966).
108. A. Trautmann, *The Applications of Fiber Bundles in Physics*, Lecture Notes, Kings College London, London (1967).
109. J. Madore, Phys. Rep. **75**, 125 (1981).
110. M. Daniel & C. M. Viallet, Rev. Mod. Phys. **52**, 175 (1980).
111. T. Eguchi et al., Phys. Rep . **66**, 213 (1980).
112. Dreschsler & Mayer, *Fiber Bundle Techniques in Gauge Theories*, Lecture Note in Physics, Springer, Berlin (1977).
113. L. P. Eisenhart, *Non-Riemannian Geometry*, Dover, New York, NY (2005).
114. M. Shifman, *Instantons in Gauge Theories*, Mikhail A. Shifman, World Scientific (1994).
115. D. S. Freed & K. K. Uhlenbeck, *Instantons and Four-Manifolds*, Springer (1984).
116. C. Gordon & R. Kirby (editors), *Four-Manifold Theory, Contemporary Mathematics*, American Mathematical Society, Providence, RI (1982).
117. R. Friedman & J. W. Morgan (editors), *Gauge Theory and the Topology of Four-Manifolds*, American Mathematical Society, Providence, RI (1997).
118. M. F. Atiyah, *Geometry of Yang-Mills Fields*, Scuola Normale Superiore, Pisa (1979).
119. David Finkelstein, Josef M. Jauch, Samuel Schiminovich, & David Speiser, J. Math. Phys. **3**, 207 (1962).
120. S. Adler, *Quaternion Quantum Mechanics and Quantum Fields*, Oxford University Press, Oxford (1995).
121. Alain Connes, *Non Commutative Geometry*, Academic, New York, NY (1994).
122. M. D. Maia & V. B. Bezerra, Int. J. Theor. Phys. **40**, 1283 (2001), hep-th/010710.
123. A. Jaffe & E. Witten, *Quantum YangMills Theory*, Clay Mathematics Institute, Cambridge, MA, http://www.claymath.org/millennium/
124. M. Gunaydin, J. Math. Phys. **17**, 1875 (1976).
125. M. Gunaydin & F. Gursey, J. Math. Phys. **14**, 1651 (1973).
126. A. Garrett Lisi, arXiv:0711.0770 [hep-th], (2007).
127. E. Alvarez, Rev. Mod. Phys. **61**, 561 (1989).
128. A. Palatini, Rend. Circ. Mat. Palermo, **43**, 203 (1919).
129. Hubert F. M. Goenner, *On the History of Unified Field Theories*, http://www.livingreviews.org/lrr-2004-2 (2004).
130. Y. M. Cho, PRD **14**, 3341 (1976).
131. C. Lanczos, Rev. Mod. Phys. **34**, 379 (1962).
132. M. Novello & A. L. Velloso, GRG **19**, 1251 (1987).
133. P. Dolan & C. W. Kim, Proc. R. Soc. London **447**, 577 (1994).
134. B. S. Edgar, Mod. Phys. Lett. A **9**, 479 (1994); ibid, J. Geom. & Phys. **54**, 251 (2005).
135. F. C. Mena & P. Todd, Class. Quant. Grav. **24**, 1733 (2007).
136. Y. M. Cho et al., arXiv:0911.3688v2 [gr-qc] 23 (2010).
137. T. Kawai, GRG **18**, 995 (1986)
138. E. Inonu & E. F. Wigner, Proc. Natl. Acad. Sci. **39**, 513 (1953).
139. Y. Murai, Prog. Theor. Phys. **9**, 147 (1952).
140. R. Feynman, *A New Approach to Quantum Theory*, World Scientific, Singapore (2010).
141. E. Witten, Commun. Math. Phys. **252**, 189 (2004).

142. K. Wilson, Phys. Rev. **D10**, 2445 (1974).
143. G. Modanese, Phys. Rev. **D49**, 6534 (1994).
144. A. Ashtekar, Phys. Rev. Lett. **57**, 2244 (1986).
145. Lee Smolin, *An Invitation to Loop Quantum Gravity*, In Quantum Theory and Symmetries, Cincinatti, pp. 655–682 (2003), hep-th/0408048.
146. M. Han & Y. Ma, Int. J. Mod. Phys. **D16**, 1397 (2007).
147. R. Penrose, *Angular Momentum: An Approach to Combinatorial Space-Time*, In Quantum Theory and Beyond, Ted Bastin (Ed.), Cambridge University Press, Cambridge (1971).
148. L. Schlaefli, Ann. Mat. **5**, 170 (1871).
149. E. Cartan, An. Soc. Polon. Math. **6**, 1 (1927).
150. M. Janet, Ann. Soc. Polon. Math. **5**, 38 (1926).
151. J. Nash, Ann. Maths. **63**, 20 (1956).
152. R. Greene, Mem. Am. Math. Soc. no. 97 (1970).
153. R. Hamilton, J. Diff. Geom **17**, 255306, (1982).
154. G. Perelman, arXiv:math/0211159.
155. J. W. York, Phys. Rev. Lett. **26**, 1656 (1971).
156. N. Arkani-Hamed et al., Phys. Lett. **B429**, 263 (1998); Phys. Rev. Lett. **84**, 586 (2000).
157. Brice deWitt, *Dynamical Theory of Groups and Fields*, In Les Houces Summer Scholl, B. S DeWitt, & C. DeWitt (Eds.), Gordon and Breach, New York, NY (1963).
158. E. Witten, Nuclear Phys. **B186**, 412 (1981).
159. M. D. Maia & W. Mecklenburg, J. Math. Phys. **25**, 3047 (1984).
160. M. D. Maia, Phys. Rev. **D31**, 262 (1985) Phys. Rev. **D31**, 268 (1985), doii:10.1103/PhysRevD.31.268.

Index

A
Abelian Groups, 25
Absolute time, 57
Action, 74
Action principle, 74
Adjoint representation, 4, 131
Affine connection, 2, 125
Affine geometry, 18
Algebra of observables, 43
Atlas, 11

B
Base manifold, 14
Base morphism, 127
Borrowed topology, 9
Bosons, 73
Bradwardine, 1
Brauer-Weyl representation, 104

C
Cartesian product, 48
Casimir operators, 41
Charts, 11
Christoffel symbols, 20
Clifford algebra, 100
Configuration space, 15
Conformal group, 162
Conservative systems, 76
Continuous groups, 26, 29
Contravariant tensors, 50
Coordinate basis, 14
Coordinate curve, 13
Coordinate transformations, 27, 107
Cossets, 26
Cotangent bundle, 44, 47
Cotangent vector field, 47
Covariant derivative, 20
Covariant tensors, 50
Curvature two-form, 140
Curves on manifolds, 13

D
Deformation tensor, 169
Deformed Riemannian Manifold, 167
Derivative map, 47
Diffeomorphism invariance, 70
Differentiable Manifold, 11
Differentiable map, 12
Differential form, 45
Dirac equation, 104
Directional derivative, 14
Distant simultaneity, 58
Dual derivative map, 47
Dual tangent bundle, 44
Dual tangent space, 44

E
Egregium theorem, 22
Einstein tensor, 68
Einstein-Hilbert Action, 158
Electromagnetic gauge group, 92
Electromagnetic Lagrangian, 95
Electroweak, 149
Electroweak unification, 151
Elementary particles, 10
Equivalent trivializations, 129
Eugene Wigner, 41
Exotic matter, 1
Exterior covariant derivative, 137
Exterior derivative, 55
Exterior product, 53
Extrinsic curvature, 21

F
Faithful representation, 29
Fermions, 73
Fiber, 14

Fiber bundle, 125
Fiber bundle morphisms, 127
Field basis, 20
Field momentum, 75
Foliation, 58
Free fall, 61
Fritz London, 3
Functional, 75
Functional variation, 108
Fundamental interactions, 1

G
Galilean group, 60
Galilean space-time, 57
Gauge curvature operator, 141
Gauge field, 142
Gauge forces, 4
Gauge interactions, 2
Gaussian curvature, 22
Gell-Mann structure Constants, 155
General covariance, 69
Generalized coordinates, 15
Geodesic, 23
Geodesic coordinates, 60
Geometrical paradigm, 157
Goldstone bosons, 88
Gravitational constant, 77
Group atlas, 26
Group contraction, 162
Group of field transformations, 27
Groups, 25
Groups of transformation, 27

H
Hamiltonian, 76
Hausdorff, 12
Hermann Weyl, 2
Higgs mechanism, 7, 84

I
Instantaneous interaction, 58
Instanton, 149
Internal symmetry, 5
Invariant subgroup, 26

K
k-forms, 52
Kernel of a Representation, 29
Killing form, 41

L
Lagrangian, 15, 75
Least energy, 84
Legendre transformation, 76, 80
Lie algebra, 4, 37
Lie group, 29
Linear operators, 28
Linear representation, 28
Local isospin, 147
Local trivialization, 129
Lorentz group, 65

M
Magnetic flux, 98
Manifold, 2, 9
Mannigfaltigkeit, 9
Mass operator, 41
Maupertuis, 74
Meissner effect, 99
Metric geometry, 18
Metricity condition, 3
Michelson-Morley, 25, 93
Mixed tensors, 49
Multiple connected group, 29

N
Newtonian space–time, 61–62
Noether, 2, 4, 107
Noether quantity, 114
Normal subgroup, 26
Nucleon, 5
Null vector, 66

O
Occam, 1
Octonion algebra, 154

P
Palatini, 158
Pauli matrices, 103
Planck regime, 8
Poincaé group, 65
Point particles, 10
Principal curvatures, 22
Principal fiber bundles, 125, 130
Product bundle, 17
Projection map, 14
Pullback, 48

Q
Quantum electrodynamics, 4
Quartic potential, 83
Quaternion algebra, 147
Quotient group, 27

R
Representation space, 28
Repulsive gravitation, 1
Ricci tensor, 64

S
Scalar field, 77
Simple connected group, 29
Simultaneity section, 58
Smooth deformations, 165
Soliton, 88
Space of perceptions, 9
Spin operator, 41
Spinor fields, 100
Spinors, 103
Spontaneously broken symmetry, 84
Structure formation, 2
Subgroups, 25
Symmetry, 25

T
Tangent bundle, 14
Tangent space, 14
Tangent vector, 13
Tangent vector fields, 15
Tensor bundle, 49
Tensor of order zero, 49
Tensor product, 48
Time translation operator, 70
Topological spaces, 9
Total space, 14
Total variation, 110
Triad holonomy, 163
Trivial fiber bundle, 127
Trivial vector bundles, 17, 127
Trivializable fiber bundle, 128
Trivializationless, 123
Trivializations, 131
Twistor, 162
Two-form, 50
Typical fiber, 127

V
Variational principle, 70
Vector bundle, 17
Vladmir Fock, 3

W
Werner Heisenberg, 5
Winding number, 99
World-line, 59

Y
Yang-Mills current, 149
Yang-Mills equations, 148